AMERICAN BOOK COMPANY'S

PASSING THE
MINNESOTA
BASIC SKILLS TEST

IN

MATHEMATICS

REVISED EDITION

COLLEEN PINTOZZI

AMERICAN BOOK COMPANY
P O BOX 2638
WOODSTOCK, GEORGIA 30188-1383
TOLL FREE: 1 (888) 264-5877   PHONE: 770-928-2834   FAX 770-928-7483

ACKNOWLEDGMENTS

I am grateful to Kelly Berg for her assistance in creating the graphics for this book. I would also like to thank Erica Day, Alan Fuqua, Mary Stoddard, Callie Aubrey, and Tracy Jones for their editorial assistance.

TABLE OF CONTENTS

MINNESOTA MATHEMATICS

Preface

PASSING THE MINNESOTA BASIC SKILLS TEST IN MATHEMATICS will help students preparing for this mathematics test. This book will also assist students who have failed the mathematics test and who want to review concepts, skills, and strategies before taking the test again. **The materials in this book are based on the objectives and content descriptions of the Minnesota Basic Skills Test published by the Minnesota Department of Children, Families, and Learning.**

This book contains several sections. These sections are: 1) General information about the test; 2) A Diagnostic Test; 3) Chapters that teach the concepts and skills emphasized on the test; 4) Two Practice Tests. Answers to the tests and exercises are in a separate manual.

We welcome comments and suggestions about the book. Please contact the author at:

American Book Company
PO Box 2638
Woodstock, GA 30188-1383

Toll Free: 1 (888) 264-5877
Phone (770) 928-2834
Fax (770) 928-7483

ABOUT THE AUTHOR

Colleen Pintozzi has taught mathematics at the middle school, junior high, senior high, and adult level for 22 years. She holds a B.S. degree from Wright State University in Dayton, Ohio and has done graduate work at Wright State University, Duke University, and the University of North Carolina at Chapel Hill. She is the author of eight mathematics books including such best-sellers as *Basics Made Easy: Mathematics Review, Passing the New Alabama Graduation Exam in Mathematics, Passing the Georgia High School Graduation Test in Mathematics, Passing the TCAP Competency Test in Mathematics, Passing the Louisiana LEAP 21 Graduation Exit Exam, Passing the Indiana ISTEP+ Graduation Qualifying Exam in Mathematics, Passing the Minnesota Basic Standards Test in Mathematics,* and *Passing the Nevada High School Proficiency Exam in Mathematics.*

PREPARING FOR THE MINNESOTA BASIC SKILLS TEST IN MATHEMATICS

INTRODUCTION

If you are a student in a Minnesota school district, you must pass the **Minnesota Basic Skills Test** to receive a secondary school diploma. This test is one of three tests now required as part of the graduation standards for Minnesota high schools. The **Basic Skills Test in Mathematics** is not a graduation-level test, but it tests the minimum level of math needed to function in society.

In this book, you'll prepare for the **Minnesota Basic Skills Test in Mathematics**. The questions and answers that follow will provide you with general information about this test.

In this book, you'll take a **Diagnostic Test** to determine your strengths and areas for improvement. In the chapters, you'll learn and practice the skills and strategies that are important in preparing for this test. The last sections contain two practice tests that will provide further preparation for the actual **Basic Skills Test**.

What Is On The Minnesota Basic Skills Test in Mathematics?

For the **Minnesota Basic Skills Test,** you will answer approximately 65 multiple-choice questions based on concepts and skills in mathematics.

Why Must I Pass the Minnesota Basic Skills Test in Mathematics?

You are required to pass this exam for several reasons. First, you must pass the mathematics, reading, and writing test to receive a diploma. Second, the state of Minnesota, your future employers, and your community need an educated workforce. Thirdly, today's high school graduates will need to adapt to rapidly changing technology throughout their lives. Employees without basic mathematics skills in computation, measurement, and geometry will be unable to compete in the workplace. Without these skills, there's a great chance they will not only be unemployed but unemployable. In addition, by demonstrating your mathematics ability, you can show what you have learned in school and apply this knowledge to new situations and experiences.

When Do I Take the Minnesota Basic Skills Test in Mathematics?

During your eighth grade year, you must take the **Minnesota Basic Skills Test in Mathematics**.

How Much Time Do I Have To Take the Test?

The **Minnesota Basic Skills Test in Mathematics** is not a timed test. As a general rule, you should finish the test in 120–150 minutes, but you will be given additional time if needed.

Can I Use a Calculator on the Minnesota Basic Skills Test in Mathematics?

Yes, you may use a calculator on all sections of the Mathematics test **except** the estimation portion of the **Minnesota Basic Skills Test in Mathematics**.

What Happens If I Don't Pass the Exam?

You may take the **Minnesota Basic Skills Test in Mathematics** several times before the end of the twelfth grade. The first time that you take this exam is in the eighth grade. You will then have at least one additional testing opportunity each year. Seniors who have not passed can take the test again in the spring of their senior year.

If I Fail the Mathematics Exam, Where Can I Get Help To Pass It the Next Time?

In some school districts, you may be able to sign up for special classes. The instructors in these classes will teach you how to study and prepare for the **Minnesota Basic Skills Test in Mathematics.** You can also seek extra help in mathematics classes during your years in middle school and high school. Finally, you may work with tutors or counselors who can help you pass the **Mathematics Exam.**

How Is My Test Scored?

If you are scheduled to graduate in the year 2004 or beyond, you must correctly answer at least 75% of the questions on the **Minnesota Basic Skills Test** in order to pass. Scoring for your exam is based on eight different areas of mathematics content. These content areas are listed below:

A) Problem Solving: Whole Numbers and Fractions

B) Percents, Rates, Ratios, and Proportions

C) Number Sense

D) Estimation

E) Measurement Concepts

F) Tables and Graphs

G) Chance and Data

H) Shape and Space

TEST-TAKING TIPS

1. **Complete the chapters and practice exams in this book.** This text will review the skills needed to pass the **Minnesota Basic Skills Test in Mathematics.**

2. **Be prepared.** Get a good night's sleep the day before your exam. Eat a well-balanced meal prior to your exam, one that contains plenty of proteins and carbohydrates.

3. **Arrive early.** Allow yourself at least 15–20 minutes to find your room and get settled. Then you can relax before the exam, so you won't feel rushed.

4. **Think success.** Keep your thoughts positive. Turn negative thoughts into positive ones. Tell yourself you will do well on the exam.

5. **Practice relaxation techniques.** Some students become overly worried about exams. Before or during the test, they may perspire heavily, experience an upset stomach, or have shortness of breath. If you feel any of these symptoms, talk to a close friend or see a counselor. They will suggest ways to deal with test anxiety. **Here are some quick ways to relieve test anxiety:**

   - Imagine yourself in your most favorite place. Let yourself sit there and relax.
   - Do a body scan. Tense and relax each part of your body starting with your toes and ending with your forehead
   - Use the 3–12–6 method of relaxation when you feel stress. Inhale slowly for 3 seconds. Hold your breath for 12 seconds, and then exhale slowly for 6 seconds.

6. **Read directions carefully.** If you don't understand them, ask the proctor for further explanation before the exam starts.

7. **Use your best approach for answering the questions.** Some test-takers like to skim the questions and answers before reading the problem or passage. Others prefer to work the problem or read the passage before looking at the answers. Decide which approach works best for you.

8. **Answer each question on the exam.** Unless you are instructed not to, make sure you answer every question. If you are not sure of an answer, take an educated guess. Eliminate choices that are definitely wrong, and then choose from the remaining answers.

9. **Use your answer sheet correctly.** Make sure the number on your question matches the number on your answer sheet. In this way, you will record your answers correctly. If you need to change your answer, erase it completely. Smudges or stray marks may affect the grading of your exams, particularly if they are scored by a computer. If your answers are on a computerized grading sheet, make sure the answers are dark. The computerized scanner will skip over answers that are too light.

10. **Check your answers.** Review your exam to make sure you have chosen the best responses. Change answers only if you are sure they are wrong.

MATHEMATICS DIAGNOSTIC TEST

The following test is intended to help students identify which objectives on the **Minnesota Basic Skills Test** they need to master. This test includes sample problems similar to ones that appear on the **Minnesota Basic Skills Test**. At the end of the test, there is an evaluation chart to help students identify areas of mastery as well as areas for improvement.

General Directions

1. Read all directions carefully. When you take the **Minnesota Basic Skills Test**, the directions will be read to you. You should follow along as the directions are read.

2. The questions on this test are followed by several suggested answers. For each question, find the <u>one</u> answer that you think is the best. Then choose that answer on your answer sheet.

3. Choose only <u>one</u> answer for each question. If you change an answer, be sure to erase the first answer completely.

4. **The Minnesota Basic Skills Test is not a timed test. You will be given all the time you need.**

5. The **first 15 questions** are estimation questions and are to be answered **without** the use of a calculator. On the **Minnesota Basic Skills Test**, you will use a sticker to seal questions 1–15 after you complete those questions. Once you have completed questions 1–15, check your answers and continue with the rest of the test. Do not go back to these questions after you begin with questions 16–76.

6. Calculators are not necessary, but may be used for questions 16 through the end of the test.

FORMULA SHEET

Formulas that you may need to work questions on this test are found below.

Area of a square = s^2

Area of a rectangle = lw

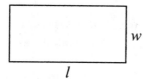

Area of a triangle = $\frac{1}{2} bh$

Area of a circle = πr^2

Circumference = πd or $2\pi r$

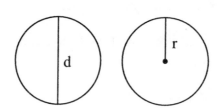

π = Pi = 3.14 or $\frac{22}{7}$

Volume of a cube = $l \times w \times h$

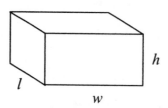

Area of a trapezoid = $\dfrac{(b_1 + b_2) \times h}{2}$

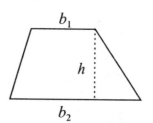

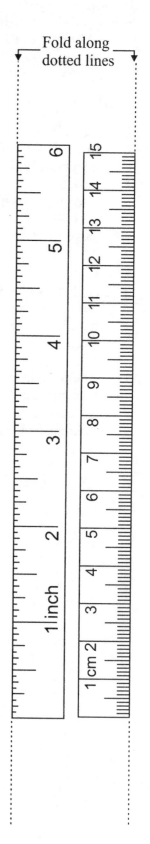

Fold along
dotted lines

x

DIAGNOSTIC TEST

1. It takes Mark 3 hours to cut the grass. Todd can cut the grass in $2\frac{1}{2}$ hours. How long will it take them if they both have a lawn mower and work together? Find the most reasonable answer.

   A. 1 hour
   B. $1\frac{1}{2}$ hours
   C. 2 hours
   D. $2\frac{3}{4}$ hours

DOBBINS FAMILY	
JANUARY	$ 89.15
FEBRUARY	$ 99.58
MARCH	$ 78.99
APRIL	$ 72.47
MAY	$ 99.23
JUNE	$124.69

2. Look at the chart above. About how much did the Dobbins family spend for electricity during the first three months of the year?

   E. $270.00
   F. $260.00
   G. $250.00
   H. $240.00

3. Pat wanted to divide 7.86 by 3.9, but he forgot to enter the decimal points when he put the numbers into the calculator. Using estimation, where should Pat put the decimal point?

   A. 0.2015386
   B. 2.015386
   C. 20.15386
   D. 201.5386

4. A piece of cloth that measures 1 yard in length is about the same length as cloth measuring _____ in length.

   E. one millimeter
   F. one decimeter
   G. one meter
   H. one kilometer

5. What is the best estimate of the weight of an average man?

   A. 5 milligrams
   B. 15 centigrams
   C. 25 grams
   D. 90 kilograms

6. Trina bought her living room furniture on the installment plan. She has 40 months to pay for the furniture interest-free. About how long does she have to pay for the furniture without interest?

    E.   2 years
    F.   3 years
    G.   4 years
    H.   40 years

LAKE HUNT

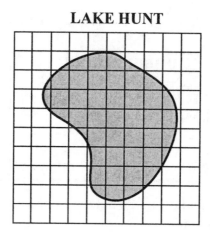

7. Lake Hunt is shown on the grid above. Each square on the grid represents 1 square mile. About how big is Lake Hunt?

    A.   30 square miles
    B.   36 square miles
    C.   45 square miles
    D.   52 square miles

8. According to the sale ad above, about how much could you save on a board game regularly priced at $12.47?

    E. $ 3.00
    F. $ 6.00
    G. $ 9.00
    H. $15.00

9. Doug earned $125 the first week on his job. The second week he earned twice as much. At the end of the second week, Doug spent $72.47. About how much money did Doug have left?

   A. $ 50.00
   B. $ 75.00
   C. $175.00
   D. $300.00

10. The regular price of a stereo is $560. The stereo is on sale for 25% off. What is the sale price of the stereo?

   E. $140
   F. $360
   G. $420
   H. $535

11. Eileen was making $8.50 per hour. Her boss gave her a $0.75 per hour raise. Eileen works 38 hours per week. What percent raise did Eileen receive?

   Which of the following is extra information not needed to answer the question?

   A. Eileen was making $8.50 per hour.
   B. Eileen got a $0.75 per hour raise.
   C. Eileen works 38 hours per week.
   D. All of the information given is needed to answer the question.

12. Right now the temperature is 5° above zero. Tonight the temperature is expected to be 12° cooler. What is the temperature expected to be?

   E. −17°
   F. −7°
   G. −12°
   H. −5°

13. At Wilkinson High School, 42% of the students ride the bus to school. There are 300 students enrolled at Wilkinson High School. How many students ride the bus to school?

   A. 126
   B. 142
   C. 174
   D. 212

14. There are 20 male and 35 female students taking band this year at Washington Middle School.   What is the ratio of female students to male students taking band this year?

    E. $\frac{4}{7}$

    F. $\frac{7}{4}$

    G. $\frac{7}{11}$

    H. $\frac{11}{7}$

15. Which set of decimals is in order from <u>LEAST</u> to <u>GREATEST</u>?

    A. .64, .604, .064, .06
    B. .603, .75, .08, .098
    C. .5, .064, .701, .901
    D. .007, .069, .69, .7

4

 If you want to check your answers to Questions 1–15, you may do so now. After you have checked your answers, continue with the rest of the test. On the Minnesota Basic Skills Test, you will seal these pages to prevent you from going back to them. You may use a calculator for the rest of the test, but you may not go back and check your answers using a calculator on Questions 1–15.

16. Which of the following sizes of dishwashing liquid costs the LEAST per ounce?

A.   4 ounces for $.68
B.   10 ounces for $1.60
C.   12 ounces for $1.92
D.   16 ounces for $2.24

17. Round 57.876 to the nearest tenth.

E.   57.880
F.   57.88
G.   57.8
H.   57.9

18. Jed has two horses, Bingo and Lightning. Lightning eats 5 kilograms of hay in 5 days. Bingo eats twice as much. How many kilograms of hay does it take to feed both horses for thirty days?

A.   5 kilograms
B.   15 kilograms
C.   30 kilograms
D.   90 kilograms

19. Julie has a part-time business selling cosmetics out of her home. This week she spent 5 hours on her business, and cosmetics sales totaled $147.50. To find out how much Julie made per hour, you also need to know

E.   how many samples she gave away.
F.   how much she spent at the grocery store.
G.   how many customers she had.
H.   how much she paid for the cosmetics and supplies.

20. It takes $1\frac{2}{3}$ yards of fabric to cover one chair. How many yards will it take to cover 6 chairs?

A. $3\frac{3}{5}$ yards
B. $4\frac{1}{3}$ yards
C. $6\frac{2}{3}$ yards
D.   10 yards

21. Which can of soup is the best buy?

    E. 39¢ per can

    F. 3 cans for 87¢

    G. 4 cans for $1.12

    H. 2 cans for 73¢

22. Mike bought iced tea for $1.05, wings for $3.45, 3 cookies for $.75 each, and paid $.34 sales tax. What was his change from $20.00?

    A. $12.91
    B. $13.41
    C. $13.96
    D. $14.41

23. The temperature right now is 7° above zero. Tonight's low is expected to be 31° colder. What is the expected low tonight?

    E. −24°
    F. 24°
    G. −31°
    H. −38°

24. A set of table and chairs that normally sells for $450.00 is on sale this week for 30% off the regular price. How much would I save if I bought it on sale?

    A. $ 30.00
    B. $ 31.50
    C. $135.00
    D. $315.00

25. If 12 cans of soup cost $5.04, how much does one can cost?

    E. 41¢
    F. 42¢
    G. 43¢
    H. 44¢

26. Todd wants a jacket that normally sells for $125.00. It is on sale for 25% off the regular price. What is the sale price of the jacket?

   A. $ 31.25
   B. $ 75.00
   C. $ 93.75
   D. $100.00

27. Billy bought the following items at the store:

1 gal milk	$2.17
2 lb hamburger	$3.98
3 donuts	$0.99
1 tube of toothpaste	$2.79
pair of socks	$1.57

   He paid 6% sales tax. What was the total cost of his purchases?

   E. $11.50
   F. $11.70
   G. $12.19
   H. $13.65

28. Ned earns $2,000 per month working on computers for IBC Corporation. His employer deducts 36% of his paycheck for taxes, insurance, and social security. How much money does Ned take home every month?

   A. $ 720
   B. $1,280
   C. $1,640
   D. $2,000

29. On a scale drawing, 3 cm represents 15 miles. A line segment on the drawing measures 9 cm. What distance is represented by the line segment?

   E. 9 miles
   F. 21 miles
   G. 27 miles
   H. 45 miles

30. Brianna makes a 20% commission on the jewelry she sells. How much commission will she make for selling $600 worth of jewelry?

   A. $ 20
   B. $120
   C. $480
   D. $620

31. Rogers' Ranch has 30 chickens, 54 cows, and 3 bulls. What is the ratio of bulls to cows?

   E. $\frac{3}{5}$

   F. $\frac{1}{18}$

   G. $\frac{1}{10}$

   H. $\frac{1}{28}$

Scale $\frac{1}{4}$ inch = 13.7 miles

32. What is the straight line distance between Duluth and Saint Cloud?

   A.   137 miles
   B.   180 miles
   C.   192 miles
   D.   207 miles

33. Which set of decimals is in order from <u>LEAST</u> to <u>GREATEST</u>?

   E.   .23, .203, .213, .3
   F.   .3, .23, .203, .213
   G.   .23, .3, .203, .213
   H.   .203, .213, .23, .3

34. $\frac{7}{8}$ of all the graduating seniors have signed up to go to the graduation dance. What percent of the seniors are going to the dance?

    A.   0.78%
    B.   0.875%
    C.   8.75%
    D.   87.5%

35. 40% of the junior class ride the bus to school. What fractional portion is this?

    E. $\frac{4}{100}$
    F. $\frac{1}{8}$
    G. $\frac{1}{4}$
    H. $\frac{2}{5}$

36. Emily measured a line segment with a ruler. The length was between $\frac{3}{8}$ and $\frac{1}{2}$ inch. What could the length have been?

    A. $\frac{5}{8}$ inch
    B. $\frac{1}{4}$ inch
    C. $\frac{7}{16}$ inch
    D. $\frac{3}{4}$ inch

37. Following the correct order of operations, what is the solution to:

$$16 + 5 - 12 \div 3 =$$

    E.   3
    F.   17
    G.   −3
    H.   6

38. Which of the following is equivalent to 2^3?

    A.   5
    B.   6
    C.   8
    D.   12

39. Gail bought a coat that normally sold for $199.99. It was selling for 45% off the regular price. Her calculator showed she could save $89.9955 on the coat. How much will that be rounded off to the nearest dollar?

    E. $ 89.00
    F. $ 89.99
    G. $ 90.00
    H. $100.00

40. The Taco Palace Mexican restaurant uses 18 pounds of Monterey Jack cheese every day. The restaurant is open 7 days a week. The restaurant orders the cheese in 10-pound blocks. How many 10-pound blocks of cheese should the cooks order each week to have enough?

    A.   2
    B.   10
    C.   13
    D.   126

41. Doug drove 416 miles at an average speed of 52 miles per hour. If he started at 1:00 p.m., when did he arrive at his destination?

    E.   6:00 p.m.
    F.   7:00 p.m.
    G.   8:00 p.m.
    H.   9:00 p.m.

42. What would be the best measure to use to find the area of a school lunchroom?

    A.   square kilometers
    B.   square meters
    C.   square centimeters
    D.   square millimeters

43. What is the length of the following line segment? (Use a ruler.)

    _____

    E.   6.8 centimeters
    F.   6.8 meters
    G.   68 centimeters
    H.   68 meters

44.
    ┌─────────────────────────────┐
    │   4 minutes 30 seconds   │
    │   ×   5   │
    │   ─────────────────────   │
    └─────────────────────────────┘

    A.   4 minutes 150 seconds
    B.   20 minutes 35 seconds
    C.   22 minutes 30 seconds
    D.   22 minutes 150 seconds

45. Andrea started driving at 5:30 a.m. After $4\frac{2}{3}$ hours on the road, she stopped for a snack. What time was it then?

    E.   10:30 a.m.
    F.   10:10 a.m.
    G.   9:50 a.m.
    H.   9:40 a.m.

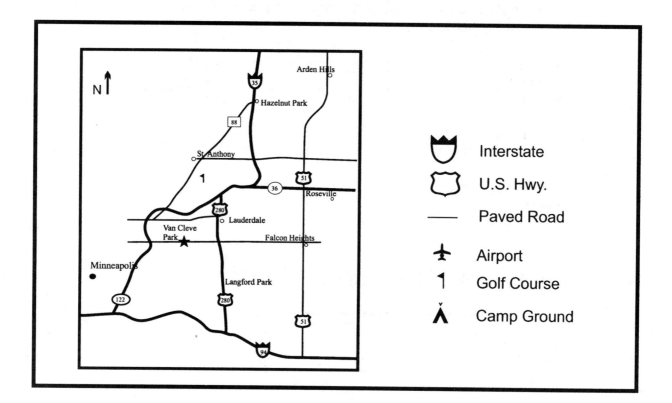

46. If you drove from Falcon Heights to Hazelnut Park, which highways would you use for the shortest route?

    A.   280, 36 and 35
    B.   88, 51 and 35
    C.   51, 36, and 35
    D.   94, 280 and 88

Use the protractor below to answer the following question.

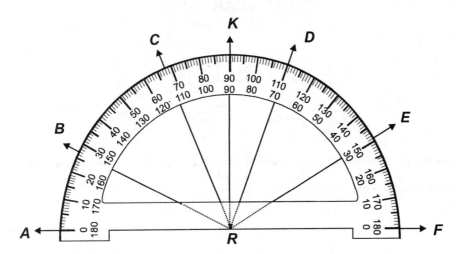

47. What is the measure of ∠*BRF*?

    E.   27°
    F.   33°
    G.   153°
    H.   167°

48. Using your ruler, find the perimeter of the following rectangle.

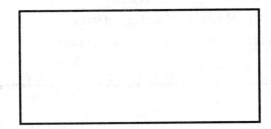

    A.   19.0 millimeters
    B.   19.0 centimeters
    C. 190.0 centimeters
    D.   1.9 meters

DUNLAP'S CAR SALES
(First 8 Months)

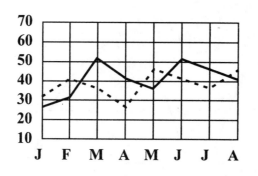

actual ⎯⎯⎯
predicted - - - - -

49. For how many months were the actual sales at Dunlap's above the predicted sales?

   E.   2
   F.   3
   G.   4
   H.   5

Sam's Seed Catalog
Purple Passion Asparagus
For each offer ordered, you get 10 plants.

Size	1 Offer	2–3 Offers	4 + Offers
1-Yr	$7.98	$7.23	$6.40
2-Yr	8.98	8.23	7.38
3-Yr	9.98	9.24	8.18

Add $2.50 to your order for shipping and handling.

50. Andy ordered 40 of the 2-yr old plants. How much money did he need to send with his order?

   A. $29.52
   B. $32.02
   C. $73.80
   D. $76.30

APPLIANCE	OHMS	VOLTS	AMPS
Coffee maker	20	110	5
Toaster	22	110	5.5
Radio	220	110	0.5
Dryer	25	220	8.8
Dishwasher	70	220	3.1

Watts = Volts × Amps

51. Using the formula and the chart above, how many watts of electrical power are required to operate a toaster?

E. 115.5
F. 121
G. 605
H. 6,050

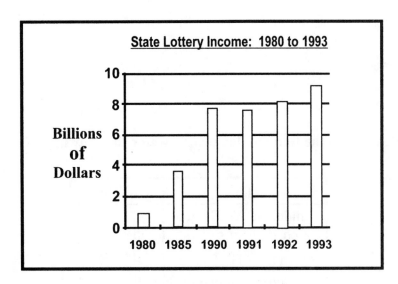

52. How much was the increase in income from 1985 to 1990?

A. 2 billion dollars
B. 3 billion dollars
C. 4 billion dollars
D. 5 billion dollars

MILEAGE CHART

	Brainerd	Hinckley	Minneapolis	Rochester	St. Paul	Worthington
Brainerd	0	99	124	213	132	237
Hinckley	99	0	79	157	74	254
Minneapolis	124	79	0	90	10	175
Rochester	213	157	90	0	83	176
St. Paul	132	74	10	83	0	180
Worthington	237	254	175	176	180	0

53. How many miles is it from Minneapolis to Rochester?

   E. 74 miles
   F. 79 miles
   G. 90 miles
   H. 213 miles

A neighborhood surveyed the times of day people water their lawns and tallied the data below.

Time	Tally
midnight – 3:59 a.m.	II
4:00 a.m. – 7:59 a.m.	JHT I
8:00 a.m. – 11:59 a.m.	JHT IIII
noon – 3:59 p.m.	JHT
4:00 p.m. – 7:59 p.m.	JHT JHT
8:00 p.m. – 11:59 p.m.	JHT III

54. Which time of day is the most popular for the people in this neighborhood to water their lawns?

   A. 4:00 a.m. – 7:59 a.m.
   B. 8:00 a.m. – 11:59 a.m.
   C. 4:00 p.m. – 7:59 p.m.
   D. 8:00 p.m. – 11:59 p.m.

Lane Publishing Company
2003 Total Sales

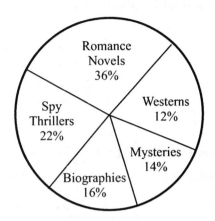

55. According to the graph, which kind of books sold the most copies in 2003?

    E.   Biographies
    F.   Mysteries
    G.   Spy Thrillers
    H.   Romance Novels

The student council surveyed the student body on favorite lunch items. The frequency chart below shows the results of the survey.

Favorite Lunch Item	Frequency
corndog	140
hamburger	245
hotdog	210
pizza	235
spaghetti	90
other	65

56. Based on the data in the chart, a student chosen at random is most likely to want which two choices for lunch?

    A.   a corndog or pizza
    B.   a hotdog or spaghetti
    C.   a hamburger or pizza
    D.   a hotdog or pizza

57. David owns a dog named Wishes. He reached into his box of 8 lamb, 12 liver, 6 chicken-flavored, and 20 milk-flavored dog biscuits and gave one to Wishes without looking. What is the probability that Wishes got a liver treat?

   E. $\frac{1}{6}$

   F. $\frac{6}{17}$

   G. $\frac{6}{23}$

   H. $\frac{1}{23}$

Use the following spinner for questions 58 and 59.

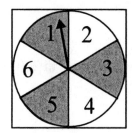

58. What is the probability that the spinner above will land on an even-numbered section?

   A. $\frac{1}{6}$

   B. $\frac{1}{2}$

   C. $\frac{1}{3}$

   D. $\frac{2}{3}$

59. The spinner above stopped on the number 5 on the first spin. What is the probability that it will not stop on the number 5 on the second spin?

   E. $\frac{1}{5}$

   F. $\frac{1}{3}$

   G. $\frac{1}{6}$

   H. $\frac{5}{6}$

60. Rami has an aquarium with 3 black goldfish and 4 orange goldfish. He purchased 2 more black goldfish to add to his aquarium. What is the new ratio of black goldfish to total goldfish?

   A. $\frac{2}{9}$

   B. $\frac{5}{4}$

   C. $\frac{4}{5}$

   D. $\frac{5}{9}$

61. Twenty-six cards each having a different letter of the alphabet were placed face dow[n] table. Tyrone picked a card at random. What is the probability that he chose a vowe[l] E, I, O, or U)?

    E. $\frac{1}{26}$

    F. $\frac{5}{26}$

    G. $\frac{21}{26}$

    H. $\frac{1}{5}$

62. Concession stand sales for each game in the season were: $320, $540, $230, $450, $280, and $580. What is the mean sales per game?

    A. $230
    B. $350
    C. $385
    D. $400

The following are student run times for the 50 meter dash:

Nate: 3 seconds	Kate: 5 seconds	Jason: 7 seconds
Pete: 4 seconds	Wanda: 6 seconds	May: 8 seconds
Ron: 4 seconds	Adam: 7 seconds	Cy: 8 seconds

63. Which student represents the median time?

    E. Nate
    F. Kate
    G. Wanda
    H. Cy

64. Tom's math grades so far this semester have been 94, 72, 77, 76, and 41. What is the mean of his grades?

    A. 53
    B. 72
    C. 86
    D. 94

65. Find the volume of the following:

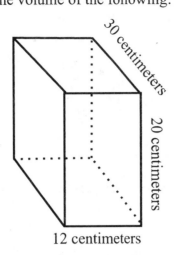

30 centimeters

20 centimeters

12 centimeters

    E.   62 cubic centimeters
    F.   612 cubic centimeters
    G.   720 cubic centimeters
    H. 7,200 cubic centimeters

66. Rhonda needs to build a fence for her dog. How many feet of fence does she need for a yard 100 feet wide and 120 feet long?

    A.   220 feet
    B.   440 feet
    C.   1,200 feet
    D. 12,000 feet

3 inches   3 inches   3 inches

3 inches

67. The drawing above shows a large square containing 4 smaller squares. What is the area of the larger square?

    E.   12 square inches
    F.   24 square inches
    G.   36 square inches
    F.   48 square inches

68. What is the area of the following triangle?

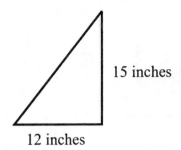

15 inches

12 inches

A. 27 square inches
B. 90 square inches
C. 135 square inches
D. 180 square inches

69. Patrick's tractor has tires with a 14-inch radius. If he drives the tractor down the road until each tire has gone around 20 times, how far has he driven?
Use $\pi = \frac{22}{7}$.

E. 44 inches
F. 880 inches
G. 1,760 inches
H. 12,320 inches

70. Tara bought a 16 inch cube for a planter. How many cubic inches could it hold?

A. 16 cubic inches
B. 48 cubic inches
C. 256 cubic inches
D. 4,096 cubic inches

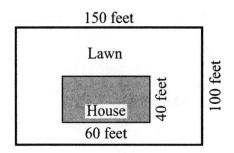

150 feet

Lawn

House

60 feet

40 feet

100 feet

71. William drew a picture of his yard to help figure the area of the yard. He needed to know how much grass seed and fertilizer to buy. The shaded area represents his house. How many square feet of yard does he have?

    E.   10,000 square feet
    F.   12,600 square feet
    G.   14,760 square feet
    H.   15,000 square feet

72. Which line segments are perpendicular in the figure below?

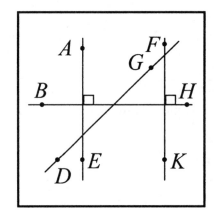

    A. $\overline{AE}$ and $\overline{BH}$
    B. $\overline{GD}$ and $\overline{FK}$
    C. $\overline{AE}$ and $\overline{GD}$
    D. $\overline{AE}$ and $\overline{FK}$

73. This table is a long distance telephone rate chart. How much would a 12-minute call be to Zone 2 from 6:30 pm to 6:42 pm?

E. $ .99
F. $1.20
G. $ .84
H. $1.12

Time	Zone 1	Zone 2
7 am to 6 pm	$.09/min	$.12/min
6 pm to 9 pm	$.07/min	$.10/min
9 pm to 7 am	$.05/min	$.07/min
Holidays	$.05/min	$.07/min

74. Rachel kept track of how many scoops she sold of the five most popular flavors in her ice cream shop.

Flavor	Scoops
Vanilla Bean	30
Chunky Chocolate	36
Strawberry Coconut	44
Chocolate Peanut Butter	46
Mint Chocolate Chip	28

Which flavor of ice cream is closest to the mean number?

A. Vanilla Bean
B. Chocolate Peanut Butter
C. Chunky Chocolate
D. Strawberry Coconut

75. If Brianna spins the spinner pictured to the right, which of the following is the **most** likely to happen?

E. It will land on an even number.
F. It will land on an odd shaded number.
G. It will land on an even shaded number.
H. It will land on a shaded number.

76. Sabina wants to cover a box on all six sides with white satin. The box is 4"×6"×10". If she glues the fabric on so it does not overlap, how many square inches of fabric will she use?

A. 120 square inches
B. 124 square inches
C. 240 square inches
D. 248 square inches

EVALUATION CHART
DIAGNOSTIC MATHEMATICS TEST

Directions: On the following chart, circle the question numbers that you answered incorrectly, and evaluate the results. Then turn to the appropriate topics (listed by chapters), read the explanations, and complete the exercises. Review the other chapters as needed. Finally, complete the Practice Mathematics Tests to further prepare yourself for the **Minnesota Basic Skills Test.**

		QUESTIONS	PAGES
Chapter 1:	**Whole Numbers**	18, 39	25–38
Chapter 2:	**Customary Measurements**	36, 43	39–43
Chapter 3:	**Fractions**	1, 8, 20	44–63
Chapter 4:	**Ratios, Probability, Proportions, and Scale Drawings**	14, 29, 31, 32, 56 57, 58, 59, 60, 61, 75	64–72
Chapter 5:	**Time Problems**	6, 41, 44, 45	73–82
Chapter 6:	**Decimals**	3, 9, 15, 16, 17, 21 22, 25, 33, 73	83–109
Chapter 7:	**Exponents and the Metric System**	4, 5, 38, 40, 42	110–117
Chapter 8:	**Percents**	10, 13, 24, 26, 27, 28, 30, 34, 35	118–132
Chapter 9:	**Problem-Solving and Critical Thinking**	8, 11, 19, 40	133–153
Chapter 10:	**Mean and Median**	62, 63, 64, 74	154–158
Chapter 11:	**Data Interpretation**	2, 46, 49, 50, 51, 52, 53, 54, 55, 73	159–173
Chapter 12:	**Integers and Order of Operations**	12, 23, 37	174–182
Chapter 13:	**Plane Geometry**	7, 48, 66, 67, 68, 68, 71, 72	183–205
Chapter 14:	**Angles**	47	206–213
Chapter 15:	**Solid Geometry**	65, 70, 76	214–228

WHOLE NUMBERS

ADDING WHOLE NUMBERS

EXAMPLE: Find $302 + 54 + 712 + 9$

Step 1: Remember when you add to arrange the numbers in columns with the ones digits at the right.

Step 2: Start at the right and add each column. Remember to carry when necessary.

$$
\begin{array}{r}
302 \\
54 \\
712 \\
+\ 9 \\
\end{array}
$$

$$
\begin{array}{r}
1 \\
302 \\
54 \\
712 \\
+\ 9 \\
\hline
1{,}077 \\
\end{array}
$$

Find the sum, and circle your answer.

1. $18 + 24 + 157$

2. $2{,}458 + 5{,}011$

3. $4{,}005 + 1{,}342$

4. $386 + 54 + 3$

5. $4{,}057 + 21 + 219$

6. $2{,}465 + 486$

7.   The total of 9 and 104

8.   94 more than 541

9.   784 increased by 51

10. 18 more than 149

11. 5 more than 557

12. 102 added to 73

13. 298 increased by 25

14. 541 plus 402

15. $12 + 454 + 3 + 97$

16. The sum of 308 and 52

17. The total of 85, 78, and 215

18. $6 + 243 + 19$

SUBTRACTING WHOLE NUMBERS

EXAMPLE: Find 1006 − 568

Step 1: Remember when you subtract to arrange the numbers in columns with the ones digits at the right.

$$\begin{array}{r} 1006 \\ -\ 568 \end{array}$$

Step 2: Start at the right, and subtract each column. Remember to borrow when necessary.

$$\begin{array}{r} 9\ 9 \\ \cancel{100}6 \\ -\ 5\ 6\ 8 \\ \hline 4\ 3\ 8 \end{array}$$ ← Borrow 1 from the 100, making it 99.

Note: **When you see "less than" in a problem, the second number becomes the top number when you set up the problem.**

Find the difference, and circle your answer.

1. 541 − 35

2. 6007 − 279

3. 694 − 287

4. 902 − 471

5. 500 − 376

6. 1047 − 483

7. 14 less than 607

8. 881 decreased by 354

9. The difference between 384 and 29

10. 560 decreased by 125

11. 43 less than 752

12. 74 less than 1093

13. 96 less than 704

14. 327 less than 1002

15. The difference between 273 and 55

16. The difference between 2849 and 756

17. 975 decreased by 249

18. 405 decreased by 36

SUBTRACTING - BORROWING TWICE

EXAMPLE: Find 4034 − 365

Step 1: Arrange the numbers in columns
with the ones digits at the right.

$$\begin{array}{r} 4034 \\ -\ 365 \end{array}$$

Step 2: Start at the right, and subtract each column.
Remember to borrow when necessary.

$$\begin{array}{r} 2\,^1 \\ 40\cancel{3}4 \\ -\ 365 \\ \hline 9 \end{array}$$

Borrow 1 from 3.
You cannot take 6 from 2 so...

Step 3:

$$\begin{array}{r} 39\,^12 \\ \cancel{40\cancel{3}}^14 \\ -\ 365 \\ \hline 3669 \end{array}$$

Borrow 1 from 40.
Now finish subtracting.

Find the difference, and check by adding. When you add, the answer should be the same as the top number of the problem.

1. $\begin{array}{r} 9\,^13 \\ \cancel{10}\cancel{4}6 \\ -678 \\ \hline 368 \end{array}$ ⟩ add to check

2. $\begin{array}{r} 3186 \\ -395 \end{array}$

3. $\begin{array}{r} 6418 \\ -3524 \end{array}$

4. $\begin{array}{r} 7416 \\ -846 \end{array}$

5. $\begin{array}{r} 5417 \\ -583 \end{array}$

6. $\begin{array}{r} 5442 \\ -679 \end{array}$

7. $\begin{array}{r} 4379 \\ -889 \end{array}$

8. $\begin{array}{r} 5462 \\ -4279 \end{array}$

9. $\begin{array}{r} 3845 \\ -1174 \end{array}$

10. $\begin{array}{r} 1724 \\ -576 \end{array}$

11. $\begin{array}{r} 4043 \\ -2995 \end{array}$

12. $\begin{array}{r} 2262 \\ -187 \end{array}$

13. $\begin{array}{r} 3784 \\ -1285 \end{array}$

14. $\begin{array}{r} 6175 \\ -482 \end{array}$

15. $\begin{array}{r} 2547 \\ -1389 \end{array}$

16. $\begin{array}{r} 1524 \\ -847 \end{array}$

17. $\begin{array}{r} 9431 \\ -5841 \end{array}$

18. $\begin{array}{r} 6703 \\ -505 \end{array}$

19. $\begin{array}{r} 8674 \\ -4883 \end{array}$

20. $\begin{array}{r} 7503 \\ -871 \end{array}$

MULTIPLYING WHOLE NUMBERS

EXAMPLE: Multiply 256 × 73

Step 1: Line up the ones digits
Multiply 256 × 3.

$$\begin{array}{r} 1 \\ 256 \\ \times\ 7\boxed{3} \\ \hline 768 \end{array}$$

Step 2: Multiply 256 × 7.
Remember to shift the
product one place to the
left. Then add.

$$\begin{array}{r} 256 \\ \times\ \boxed{7}3 \\ \hline 768 \\ 1792 \\ \hline 18,688 \end{array}$$

Multiply.

1. $\begin{array}{r} 258 \\ \times\ 72 \\ \hline \end{array}$

2. $\begin{array}{r} 742 \\ \times\ 44 \\ \hline \end{array}$

3. $\begin{array}{r} 785 \\ \times\ 32 \\ \hline \end{array}$

4. $\begin{array}{r} 679 \\ \times\ 36 \\ \hline \end{array}$

5. $\begin{array}{r} 841 \\ \times\ 27 \\ \hline \end{array}$

6. $\begin{array}{r} 324 \\ \times\ 19 \\ \hline \end{array}$

7. $\begin{array}{r} 921 \\ \times\ 23 \\ \hline \end{array}$

8. $\begin{array}{r} 454 \\ \times\ 56 \\ \hline \end{array}$

9. $\begin{array}{r} 156 \\ \times\ 95 \\ \hline \end{array}$

10. $\begin{array}{r} 765 \\ \times\ 94 \\ \hline \end{array}$

11. $\begin{array}{r} 581 \\ \times\ 25 \\ \hline \end{array}$

12. $\begin{array}{r} 827 \\ \times\ 56 \\ \hline \end{array}$

13. $\begin{array}{r} 942 \\ \times\ 24 \\ \hline \end{array}$

14. $\begin{array}{r} 247 \\ \times\ 84 \\ \hline \end{array}$

15. $\begin{array}{r} 468 \\ \times\ 43 \\ \hline \end{array}$

16. $\begin{array}{r} 456 \\ \times\ 47 \\ \hline \end{array}$

17. $\begin{array}{r} 743 \\ \times\ 65 \\ \hline \end{array}$

18. $\begin{array}{r} 527 \\ \times\ 38 \\ \hline \end{array}$

19. $\begin{array}{r} 524 \\ \times\ 39 \\ \hline \end{array}$

20. $\begin{array}{r} 682 \\ \times\ 64 \\ \hline \end{array}$

DIVIDING WHOLE NUMBERS

EXAMPLE: Divide $4993 \div 24$

Step 1: Rewrite the problem using the symbol $\overline{)}$.

Step 2: Divide 24 into 49. Multiply 2×24, and subtract.

$$
\begin{array}{r}
2 \\
24)\overline{4993} \\
-48 \\
\hline
19
\end{array}
$$

Step 3: You will notice you **cannot** divide 24 into 19. You **must** put a 0 in the answer, and then bring down the 3.

$$
\begin{array}{r}
20 \\
24)\overline{4993} \\
-48 \\
\hline
193
\end{array}
$$

Step 4 Divide 24 into 193. Multiply 8×24, and subtract.

$$
\begin{array}{r}
208\,r\,1 \\
24)\overline{4993} \\
-48 \\
\hline
193 \\
-192 \\
\hline
1
\end{array}
$$

The answer is 208 with a remainder of 1.

Divide.

1. $6274 \div 13$

2. $2384 \div 43$

3. $12747 \div 24$

4. $5417 \div 19$

5. $8042 \div 27$

6. $9548 \div 63$

7. $6254 \div 41$

8. $4362 \div 35$

9. $4345 \div 25$

10. $15467 \div 43$

11. $7412 \div 54$

12. $9379 \div 83$

13. $9547 \div 31$

14. $7436 \div 61$

15. $5464 \div 38$

16. $23567 \div 11$

ROUNDING WHOLE NUMBERS

EXAMPLE:

Ten Thousands
Thousands
Hundreds
Tens
Ones

4 8,5 3 8

Consider the number 48,538 shown at the left with the place values labeled. To round to a given place value, first find the place in the number. Then look to the digit on the right. If the digit on the right is 5 or greater, INCREASE BY ONE the place to which the number is being rounded. All the digits to the right of the given place value become 0's. If the digit on the right is LESS THAN 5, leave that place value the same, and change the digits on the right to 0's.

EXAMPLE: Round the number 48,538 to the nearest:

ten	48,540
hundred	48,500
thousand	49,000
ten thousand	50,000

Round to the nearest ten.

1. 523 _____
2. 6,745 _____
3. 1,324 _____
4. 872 _____
5. 2,421 _____
6. 749 _____
7. 8,478 _____
8. 5,498 _____

Round to the nearest hundred.

9. 659 _____
10. 4,478 _____
11. 32,197 _____
12. 12,631 _____
13. 847 _____
14. 3,945 _____
15. 752 _____
16. 45,327 _____
17. 15,602 _____
18. 9,485 _____
19. 41,974 _____
20. 3,649 _____

Round to the nearest thousand.

21. 54,985 _____
22. 62,743 _____
23. 11,342 _____
24. 7,453 _____
25. 82,571 _____
26. 132,250 _____
27. 75,299 _____
28. 20,098 _____
29. 68,987 _____
30. 822,621 _____
31. 42,165 _____
32. 4,902 _____

Round to the nearest ten thousand.

33. 28,235 _____
34. 39,678 _____
35. 242,469 _____
36. 153,958 _____
37. 458,347 _____
38. 50,512 _____
39. 85,369 _____
40. 471,963 _____

Round to the nearest dollar.

41. $18.52 _____
42. $16.23 _____
43. $114.99 _____
44. $18.13 _____

Round to the nearest ten dollars.

45. $19.56 _____
46. $48.63 _____
47. $64.25 _____
48. $84.95 _____

ESTIMATED SOLUTIONS

In the real world, estimates can be very useful. The best approach to finding estimates is to round off all numbers in the problem. Then solve the problem, and choose the closest answer. If money problems have both dollars and cents, round to the nearest dollar or ten dollars. $44.86 rounds to $40.

EXAMPLE: Which is a reasonable answer? $1580 \div 21$

A. 80   B. 800   C. 880   D. 8,000

Step 1:   Round off the numbers in the problem.   1580 rounds to 1600   21 rounds to 20

Step 2:   Work the problem. $1600 \div 20 = 80$. The closest answer is **A.** 80.

Choose the best answer below.

1. Which is a reasonable answer? 544×12 A. 54   B. 500   C. 540   D. 5400

2. Jeff bought a pair of pants for $45.95, a belt for $12.97, and a dress shirt for $24.87. Estimate about how much he spent.   A. $60   B. $70   C. $80   D. $100

3. For lunch, Marcia ate a sandwich with 187 calories, a glass of skim milk with 121 calories, and 2 brownies with 102 calories each. About how many calories did she consume?
A. 300   B. 350   C. 480   D. 510

4. Which is a reasonable answer? $89,990 \div 28$
A. 300   B. 500   C. 1,000   D. 3,000

5. Which is a reasonable answer? $74,295 - 62,304$
A. 12,000   B. 11,000   C. 10,000   D. 1,000

6. Delia bought 4 cans of soup at 99¢ each, a box of cereal for $4.78, and 2 frozen dinners at $3.89 each. About how much did she spend?
A. $10.00   B. $11.00   C. $13.00   D. $17.00

7. Which is the best estimate? $22,480 + 5,516$
A. 2,800   B. 17,000   C. 28,000   D. 32,000

School Store Price List

Pencils	Erasers	Folders	Binders	Compass	Protractor	Paper	Pens
2 for 78¢	59¢	21¢	$2.79	$1.59	89¢	$1.29	$1.10

8. Jake needs 2 pencils, 3 erasers, a binder, and a compass. About how much money will he need according to the chart?   A. $7.00   B. $8.00   C. $9.00   D. $10.00

9. Tracy needs a pack of paper, 2 folders, a protractor, and 6 pencils. About how much money does she need?   A. $3.00   B. $4.00   C. $5.00   D. $6.00

10. Which is the best estimate? $23895 \div 599$ A. 4   B. 40   C. 400   D. 4,000

TIME OF TRAVEL

EXAMPLE: Katrina drove 384 miles at an average of 64 miles per hour. How many hours did she travel?

Solution: Divide the number of miles by the miles per hour. $$\frac{384 \text{ miles}}{64 \text{ miles/hour}} = 6 \text{ hours}$$

Find the hours of travel in each problem below.

1. Bobbi drove 342 miles at an average speed of 57 miles per hour. How many hours did she drive? _____

2. Jan set her speed control at 55 miles per hour and drove for 165 miles. How many hours did she drive? _____

3. John traveled 2,092 miles in a jet that flew an average of 523 miles per hour. How long was he in the air? _____

4. How long will it take a bus averaging 54 miles per hour to travel 378 miles? _____

5. Kyle drove his motorcycle in a 225-mile race, and he averaged 75 miles per hour. How long did it take for him to complete the race? _____

6. Stacy drove 576 miles at an average speed of 48 miles an hour. How many hours did she drive? _____

7. Kendra flew 250 miles in a glider and averaged 125 miles per hour in speed. How many hours did she fly? _____

8. Travis traveled 496 miles at an average speed of 62 miles per hour. How long did he travel? _____

9. Wanda rode her bicycle an average of 15 miles an hour for 60 miles. How many hours did she ride? _____

10. Garren drove 184 miles at an average speed of 46 miles per hour. How many hours did he drive? _____

11. A train traveled at a constant 85 miles per hour for 425 miles. How many hours did the train travel? _____

12. How long was Amy on the road if she drove 195 miles at an average of 65 miles per hour? _____

RATE

EXAMPLE: Laurie traveled 312 miles in 6 hours. What was her average rate of speed?

Solution: Divide the number of miles by the number of hours. $\dfrac{312 \text{ miles}}{6 \text{ hour}} = 52$ miles/hour

Laurie's average rate of speed was 52 miles per hour (or 52 mph).

Find the average rate of speed in each problem below.

1. A race car went 500 miles in 4 hours. What was its average rate of speed?

   _____

2. Carrie drove 124 miles in 2 hours. What was her average speed?

   _____

3. After 7 hours of driving, Ed had gone 364 miles. What was his average speed?

   _____

4. Vera drove 360 miles in 8 hours. What was her average speed?

   _____

5. After 3 hours of driving, Paul had gone 183 miles. What was his average speed?

   _____

6. Sarah ran 25 miles in 5 hours. What was her average speed?

   _____

7. A train traveled 492 miles in 6 hours. What was its average rate of speed?

   _____

8. A commercial jet traveled 1,572 miles in 3 hours. What was its average speed?

   _____

9. Vanessa drove 195 miles in 3 hours. What was her average speed?

   _____

10. Greg drove 8 hours from his home to a city 336 miles away. At what average speed did he travel?

   _____

11. Michael drove 64 miles in one hour. What was his average speed in miles per hour?

   _____

12. After 9 hours of driving, Kate had traveled 405 miles. What speed did she average?

   _____

DISTANCE

EXAMPLE: Jessie traveled for 7 hours at an average rate of 58 miles per hour. How far did she travel?

Solution: Multiply the number of hours by the average rate of speed.
7 hours × 58 miles/hour = 406 miles

Find the distance in each of the following problems.

1. Myra traveled for 9 hours at an average rate of 45 miles per hour. How far did she travel?

   _____

2. A tour bus drove 4 hours averaging 58 miles per hour. How many miles did it travel?

   _____

3. Tina drove for 7 hours at an average speed of 53 miles per hour. How far did she travel?

   _____

4. Michael raced for 3 hours averaging 176 miles per hour. How many miles did he race?

   _____

5. Kris drove 5 hours and averaged 49 miles per hour. How far did she travel?

   _____

6. Oliver drove at an average of 93 miles per hour for 3 hours. How far did he travel?

   _____

7. A commercial airplane traveled 514 miles per hour for 2 hours. How far did it fly?

   _____

8. A train traveled at 125 miles per hour for 4 hours. How many miles did it travel?

   _____

9. Carmen drove a constant 65 miles an hour for 3 hours. How many miles did he drive?

   _____

10. Jasmine drove for 5 hours averaging 40 miles per hour. How many miles did she drive?

    _____

11. Roger flew his glider for 2 hours at 87 miles per hour. How many miles did his glider fly?

    _____

12. Beth traveled at a constant 65 miles per hour for 4 hours. How far did she travel?

    _____

34

MILES PER GALLON

EXAMPLE: The odometer on Ginger's car read 46,789 before she started her trip. At the end of her trip, it read 47,119. She used 10 gallons of gasoline. How many miles per gallon did she average?

Step 1: Subtract the ending odometer reading from the beginning odometer reading.
47,119 − 46,789 = 330 miles traveled

Step 2: Divide the number of miles traveled by the number of gallons of gasoline used.
$\frac{330 \text{ miles}}{10 \text{ gallons}}$ = 33 miles per gallon

Compute the number of miles per gallon in each of the following problems.

1. Rachael's odometer read 125,625 at the beginning of her trip. At the end of her trip, it read 125,863. She used 7 gallons of gasoline. How many miles per gallon did she average? _____

2. Tamera traveled 492 miles on 12 gallons of gasoline. What was her average gas mileage? _____

3. The odometer on Blake's car read 3,975 before she started her trip. At the end of her trip, it read 4,625. She used 26 gallons of gasoline. How many miles per gallon did she average? _____

4. Farmer Joe's tractor odometer read 218,754 before tilling his fields. After tilling, it read 218,802. He used 4 gallons of gasoline. How many miles per gallon did his tractor average? _____

5. When Devyn started the week, the odometer on his van read 64,742. At the end of the week, it read 64,984. He used 11 gallons of gasoline. How many miles per gallon did he average? _____

6. Kathryn drove 364 miles on 13 gallons of gasoline. What was her gas mileage? _____

7. Manny used his jeep to tow his boat to the beach. His odometer read 23,745 before his trip, and it read 24,030 once he arrived at the beach. His jeep used 15 gallons of gas. How many miles per gallon did he average? _____

8. Bonnie's odometer read 17,846 before she drove to visit her aunt. Once she arrived at her aunt's house, her odometer read 18,726. She used 22 gallons of gas. What was her average gas mileage? _____

9. Ron traveled 74 miles on 2 gallons of gas. How many miles per gallon did he average? _____

10. Before Janet and Bill left for their vacation, their odometer read 87,985. When they arrived back home, their odometer read 88,753. They used 24 gallons of gasoline. What was their average gas mileage? _____

WORD PROBLEMS

1. If Jacob averages 15 points per basketball game, how many points will he score in a season with 12 games?

   _____

2. A cashier can ring up 12 items per minute. How long will it take the cashier to ring up a customer with 72 items?

   _____

3. Mrs. Randolph has 26 students in 1st period, 32 students in 2nd period, 27 students in 3rd period, and 30 students in 4th period. What is the total number of students Mrs. Randolph teaches?

   _____

4. When Gerald started on his trip, his odometer read 109,875. At the end of his trip it read 110,480. How many miles did he travel?

   _____

5. The Beta Club is raising money by selling boxes of candy. It sold 152 boxes on Monday, 236 boxes on Tuesday, 107 boxes on Wednesday, and 93 boxes on Thursday. How many total boxes did the Beta Club sell?

   _____

6. Jonah won 1056 tickets in the arcade. He purchased a pair of binoculars for 964 tickets. How many tickets does he have left?

   _____

7. A school cafeteria has 52 tables. If each table seats 14 people, how many people can be seated in the cafeteria?

   _____

8. Leadville, Colorado is 14,286 feet above sea level. Denver, Colorado is 5,280 feet above sea level. What is the difference in elevation between these two cities?

   _____

9. The local bakery made 288 doughnuts on Friday morning. How many dozen doughnuts did they make?

   _____

10. Mattie ate 14 chocolate-covered raisins. Her big brother ate 5 times as many. How many chocolate-covered raisins did her brother eat?

    _____

11. Concession stand sales for a football game totaled $1,563. The actual cost for the food and beverages was $395. How much profit did the concession stand make?

    _____

12. An orange grove worker can harvest 480 oranges per hour by hand. How many oranges can the worker harvest in an 8 hour day?

    _____

TWO-STEP WORD PROBLEMS

1. There are 25 miniature chocolate bars in a bag. There are 20 bags in a carton. Damon needs to order 10,000 miniature chocolate bars. How many cartons will he need to order?

2. LeAnn needs 2,400 boxes for her business. The boxes she needs come in bundles of 50 that weigh 45 pounds per bundle. What will be the total weight of the 2,400 boxes she needs?

3. Seth uses 20 nails to make a birdhouse. He wants to make 60 birdhouses to sell at the county fair. There are 30 nails in a box. How many boxes will he need?

4. There are 12 computer disks in a box. There are 10 boxes in a carton. John ordered 16 cartons. How many disks is he getting?

5. The Do-Nut Factory packs 13 doughnuts in each baker's dozen box. They also sell cartons of doughnuts which have 6 baker's dozen boxes. Duncan needs to feed 780 people. Assuming each person eats only 1 doughnut, how many cartons will he need to buy from the Do-Nut Factory?

6. Brittany has 2 dogs, a Saint Bernard and a Golden Retriever. The Saint Bernard eats twice as much as the Golden Retriever. The Retriever eats 5 pounds of food in 6 days. How many pounds of food do the two dogs eat in 30 days?

7. Each of the 4 engines on a jet uses 500 gallons of fuel per hour. How many gallons of fuel are needed for a 5 hour flight with enough extra fuel for an additional 2 hours as a safety precaution?

8. The Farmer's Dairy has 1,620 pounds of butter to package. They are packaging the butter in five-pound tubs to distribute to restaurants. If they put 12 tubs in a case, how many cases of butter can they fill?

9. Tom has 155 head of cattle. Each eats 8 pounds of grain per day. How many pounds of grain does Tom need to feed his cattle for 10 days?

10. When you grind 3 cups of grain, you get 5 cups of flour. How many cups of grain must you grind to get 40 cups of flour?

CHAPTER 1 REVIEW

1. Add: $18 + 694 + 123 + 75$

2. Subtract: $943 - 768$ _____

3. Multiply: 452×23 _____

4. Divide: $786 \div 95$ _____

_____

Round to the nearest 100.

5.	215	_____	8. 351	_____
6.	553	_____	9. 929	_____
7.	345	_____	10. 872	_____

Round to the nearest 10.

11.	67	_____	14. 42	_____
12.	951	_____	15. 1259	_____
13.	683	_____	16. 563	_____

Round to the nearest dollar.

17.	$13.65	_____	20. $18.03	_____
18.	$22.12	_____	21. $20.54	_____
19.	$45.97	_____	22. $33.49	_____

Round to the nearest ten dollars.

23.	$42.97	_____	26. $56.22	_____
24.	$65.03	_____	27. $31.95	_____
25.	$184.99	_____	28. $473.82	_____

29. The animal keeper feeds Mischief, the monkey, 5 pounds of bananas per day. The gorilla eats 4 times as many bananas as the monkey. How many pounds of bananas does the animal keeper need to feed both animals for a week?

_____

30. A textile manufacturer uses 170 ft^2 of fabric to make a set of sheets. How many sets of sheets can it make from a 8,840 ft^2 roll of fabric?

_____

31. The Bing family's odometer read 65,453 before driving to Disney World for vacation. After their vacation, the odometer read 66,245. How many miles did they drive during their vacation?

_____

32. Jonathan can assemble 47 widgets per hour. How many can he assemble in an 8 hour day?

_____

33. Jacob drove 252 miles, and his average speed was 42 miles per hour. How many hours did he drive?

_____

34. The Jones family traveled 300 miles in 5 hours. What was their average speed?

_____

35. Connie drove for 2 hours at a constant speed of 55 miles per hour. How many total miles did she travel?

_____

36. Julie drove 189 miles and used 9 gallons of gas. What was her gas mileage?

_____

Chapter 2

CUSTOMARY MEASUREMENTS

USING THE RULER

Practice measuring the objects below with a ruler.

Measure these distances in inches.

1. How tall is the calculator? _____

2. How wide is the maple leaf? _____

3. How long is the car? _____

4. How far is it from the nose of the airplane to the nose of the camel? _____

5. How long is the trumpet? _____

6. How far is it from the middle of the bicycle's back wheel to the middle of the front wheel? _____

7. How long is the hour hand on the clock? _____

8. How far is it from the nose of the car to the mouthpiece of the trumpet? _____

Measure these distances in centimeters.

9. How long is the minute hand of the clock? _____

10. How tall is the maple leaf? _____

11. How tall is the camel? _____

12. How wide is the calculator? _____

13. How long is the plane? _____

14. How far is it from the tip of the hour hand to the tallest hump on the camel's back? _____

15. How far is it from the tip of the camel's nose to its front hoof? _____

16. How wide is the airplane from wing tip to wing tip? _____

MORE MEASURING

Measure the following line segments.

1. ———————————— = _____ in   6. ———————————— = _____ cm

2. ————— = _____ in   7. ——————— = _____ cm

3. ——————————— = _____ in   8. ——————————— = _____ cm

4. ————————————— = _____ in   9. ————————————————— = _____ cm

5. ——————————— = _____ in   10. ————— = _____ cm

Measure the dimensions of the following figures.

11.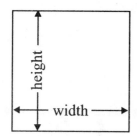

height: _____ cm

width: _____ cm

12.

height from the top of the
stem to the bottom:

_____ in

13.

width of the desk top:

_____ in

height of the desk:

_____ in

14.

length of guitar:

_____ cm

15.

width of soccer ball:

_____ cm

16.

height of tent: _____ in

width of tent at its base:

_____ in

40

APPROPRIATE INSTRUMENTS AND UNITS OF MEASURE

English System of Measure

Measure	Abbreviations	Appropriate Instrument
Time: 1 week = 7 days 1 day = 24 hours 1 hour = 60 minutes 1 minute = 60 seconds	week = wk hour = hr or h minutes = min seconds = sec	calendar clock clock clock
Length: 1 mile = 5,280 feet 1 yard = 3 feet 1 foot = 12 inches	mile = mi yard = yd foot = ft inch = in	odometer yard stick, tape line ruler, yard stick
Volume: 1 gallon = 4 quarts 1 quart = 2 pints 1 pint = 2 cups 1 cup = 8 ounces	gallon = gal quart = qt pint = pt ounce = oz	quart or gallon container quart container cup, pint, or quart container cup
Weight: 16 ounces = 1 pound	pound = lb ounce = oz	scale or balance

Look at the chart above to help you determine which instrument you would use to measure each item below.

Item: Instrument:

1.   How much water a pan holds   _____

2.   Weight of a crate of apples   _____

3.   Distance from Atlanta to Minneapolis   _____

4.   How much water a glass holds   _____

5.   Width of a room   _____

6.   Length of a shoe string   _____

7.   How long it takes to run a mile   _____

8.   How many days until school gets out   _____

9.   How deep a snow drift is   _____

10.   How many hours you were at the mall   _____

11.   Height of a woman   _____

12.   Length of a necklace   _____

Matching

Match the item on the left with its approximate (not exact) measure on the right. You may use some answers more than once.

_____   1. The height of an average woman is about _____ .

_____   2. An average candy bar weighs about _____ .

_____   3. An average doughnut is about _____ across.

_____   4. A piece of notebook paper is about _____ long.

_____   5. The average snowball is about _____ across.

_____   6. The average basketball is about _____ across.

_____   7. The average month is about _____ .

_____   8. How long is the average lunch table?

_____   9. About how much does a computer disk weigh?

_____   10. What is the average height of a table?

A. 1 yard

B. 2 yards

C. $5\frac{1}{2}$ feet

D. 4 weeks

E. 3 inches

F. 2 ounces

G. 1 foot

SIMPLIFYING UNITS OF MEASURE

EXAMPLE: Simplify: 12 pounds 35 ounces

Step 1: 35 ounces is more than 1 pound. There are 16 ounces in a pound, so divide 35 by 16.

```
        2 lb
   16) 3 5
      - 3 2
        3 oz
```

        12 pounds 35 ounces
    +   2 pounds   3 ounces
        14 pounds   3 ounces

Simplify the following.

1. 3 pounds 20 ounces

2. 2 cups 12 ounces

3. 2 yards 38 inches

4. 1 pint 1 cup 16 ounces

5. 1 yard 7 feet 12 inches

6. 1 gallon 6 quarts 3 pints

7. 3 yards 10 feet 18 inches

8. 6 gallons 4 quarts

9. 2 feet 18 inches

10. 1 pound 33 ounces

11. 6 yards 1 foot 15 inches

12. 3 cups 20 ounces

CHAPTER 2 REVIEW

Measure the following distances.

1. The length of the pencil _____ inches

2. The width of the bell _____ inches

3. The width of the pencil _____ centimeters

4. The distance between the point of the pencil and the top of the sail _____ centimeters

Measure the following line segments.

5. ———————————— _____ in   7. ——— _____ in

6. ———————————— _____ cm   8. ————— _____ cm

Measure the dimensions of the following figures.

9.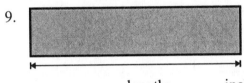

length: _____ inches

11.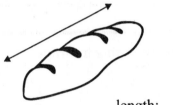

length: _____ cm

10.

height: _____ inches

12.

height: _____ cm

Simplify the following problems.

13. 2 feet 14 inches

14. 6 quarts 20 ounces

15. 5 gallons 1 quarts 4 pints

16. 5 yards 11 feet 7 inches

17. 5 gallons 2 quarts 11 cups

18. 13 pints 11 ounces

19. 3 pounds 25 ounces

20. 5 cups 35 ounces

21. 4 pounds 19 ounces

FRACTIONS

SIMPLIFYING IMPROPER FRACTIONS

EXAMPLE: Simplify: $\dfrac{11}{6}$

Step 1: $\dfrac{11}{6}$ is the same as $11 \div 6$. $11 \div 6 = 1$ with a remainder of 5.

Step 2: Rewrite as a whole number with a fraction: $1\dfrac{5}{6}$

EXAMPLE: Simplify: $\dfrac{21}{4} = 21 \div 4 = 5$ remainder 1

$\dfrac{21}{4} \longrightarrow 5\dfrac{1}{4}$ The bottom number of the fraction always remains the same.

Simplify the following improper fractions.

1. $\dfrac{11}{4} =$ ____   4. $\dfrac{6}{5} =$ ____   7. $\dfrac{12}{5} =$ ____   10. $\dfrac{7}{3} =$ ____   13. $\dfrac{8}{5} =$ ____   16. $\dfrac{9}{4} =$ ____

2. $\dfrac{10}{3} =$ ____   5. $\dfrac{17}{4} =$ ____   8. $\dfrac{19}{4} =$ ____   11. $\dfrac{13}{10} =$ ____   14. $\dfrac{14}{5} =$ ____   17. $\dfrac{15}{3} =$ ____

3. $\dfrac{20}{4} =$ ____   6. $\dfrac{12}{3} =$ ____   9. $\dfrac{25}{6} =$ ____   12. $\dfrac{31}{8} =$ ____   15. $\dfrac{13}{12} =$ ____   18. $\dfrac{24}{6} =$ ____

Add and simplify. The first one is done for you.

19. $\dfrac{3}{8}$ 20. $\dfrac{4}{5}$ 21. $\dfrac{7}{16}$ 22. $\dfrac{7}{10}$ 23. $\dfrac{5}{12}$ 24. $\dfrac{7}{8}$ 25. $\dfrac{5}{6}$ 26. $\dfrac{1}{4}$ 27. $\dfrac{1}{12}$

$\dfrac{5}{8}$ $\dfrac{2}{5}$ $\dfrac{9}{16}$ $\dfrac{3}{10}$ $\dfrac{7}{12}$ $\dfrac{7}{8}$ $\dfrac{5}{6}$ $\dfrac{3}{4}$ $\dfrac{5}{12}$

$+ \dfrac{1}{8}$ $+ \dfrac{3}{5}$ $+ \dfrac{5}{16}$ $+ \dfrac{9}{10}$ $+ \dfrac{11}{12}$ $+ \dfrac{5}{8}$ $+ \dfrac{1}{6}$ $+ \dfrac{3}{4}$ $+ \dfrac{7}{12}$

$\dfrac{9}{8} = 1\dfrac{1}{8}$

CHANGE MIXED NUMBERS TO IMPROPER FRACTIONS

EXAMPLE: Change $4\frac{3}{5}$ to an improper fraction.

Step 1: Multiply the whole number (4) by the bottom number of the fraction (5). $4 \times 5 = 20$

Step 2: Add the top number. $20 + 3 = 23$

Step 3: Put the answer over the bottom number (5).

3. Put the answer here.

2. Add this number. ⟶ $4\frac{3}{5} = \frac{23}{5}$

1. Multiply these two numbers.

4. This number stays the same.

Notice the bottom number stays the same.

Change the following mixed numbers to improper fractions.

1. $4\frac{3}{5} =$ _____

2. $6\frac{7}{8} =$ _____

3. $5\frac{3}{10} =$ _____

4. $10\frac{1}{5} =$ _____

5. $7\frac{2}{5} =$ _____

6. $1\frac{4}{5} =$ _____

7. $4\frac{9}{10} =$ _____

8. $8\frac{2}{3} =$ _____

9. $2\frac{1}{12} =$ _____

10. $9\frac{1}{3} =$ _____

11. $7\frac{7}{8} =$ _____

12. $2\frac{3}{4} =$ _____

13. $5\frac{5}{6} =$ _____

14. $1\frac{5}{8} =$ _____

15. $3\frac{3}{8} =$ _____

16. $3\frac{4}{5} =$ _____

17. $6\frac{3}{5} =$ _____

18. $6\frac{1}{8} =$ _____

19. $9\frac{1}{6} =$ _____

20. $2\frac{7}{10} =$ _____

Whole numbers become improper fractions when you put them over 1. The first one is done for you.

21. $4 = \frac{4}{1}$

22. $5 =$ _____

23. $14 =$ _____

24. $2 =$ _____

25. $1 =$ _____

26. $3 =$ _____

27. $8 =$ _____

28. $4 =$ _____

29. $10 =$ _____

30. $12 =$ _____

REDUCING PROPER FRACTIONS

EXAMPLE: Reduce $\frac{4}{8}$ to lowest terms.

Step 1: Think: What is the largest number that can be divided into 4 and 8 without a remainder?

These must be the same number. $\longleftarrow$ $\begin{array}{c} ?\overline{)4} \\ ?\overline{)8} \end{array}$

Step 2: Divide the top and bottom of the fraction by the same number. $\quad \dfrac{4 \div 4}{8 \div 4} = \dfrac{1}{2}$

Reduce the following fractions to lowest terms.

1. $\dfrac{7}{14} =$ 7. $\dfrac{6}{15} =$ 13. $\dfrac{10}{24} =$ 19. $\dfrac{15}{21} =$ 25. $\dfrac{7}{21} =$

2. $\dfrac{4}{12} =$ 8. $\dfrac{4}{14} =$ 14. $\dfrac{12}{30} =$ 20. $\dfrac{2}{6} =$ 26. $\dfrac{30}{40} =$

3. $\dfrac{8}{12} =$ 9. $\dfrac{6}{27} =$ 15. $\dfrac{4}{6} =$ 21. $\dfrac{14}{42} =$ 27. $\dfrac{12}{36} =$

4. $\dfrac{5}{20} =$ 10. $\dfrac{3}{9} =$ 16. $\dfrac{14}{35} =$ 22. $\dfrac{16}{36} =$ 28. $\dfrac{8}{18} =$

5. $\dfrac{2}{8} =$ 11. $\dfrac{7}{21} =$ 17. $\dfrac{42}{70} =$ 23. $\dfrac{18}{30} =$ 29. $\dfrac{30}{50} =$

6. $\dfrac{8}{20} =$ 12. $\dfrac{8}{28} =$ 18. $\dfrac{12}{18} =$ 24. $\dfrac{10}{32} =$ 30. $\dfrac{8}{10} =$

MULTIPLYING FRACTIONS

EXAMPLE: Multiply $4\frac{3}{8} \times \frac{8}{10}$

Step 1: Change the mixed numbers in the problem to improper fractions. $\quad \frac{35}{8} \times \frac{8}{10}$

Step 2: When multiplying fractions, you can cancel and simplify terms that have a common factor. To cancel 2 numbers, one number must be on the top (a numerator) and the other on the bottom (a denominator).

The 8 in the first fraction will cancel with the 8 in the second fraction.

The terms 35 and 10 are both divisible by 5, so 35 simplifies to 7, and 10 simplifies to 2.

$$\frac{\overset{7}{\cancel{35}}}{\cancel{8}} \times \frac{\cancel{8}^{1}}{\cancel{10}_{2}}$$

Step 3: Multiply the simplified fractions. $\quad \frac{7}{1} \times \frac{1}{2} = \frac{7}{2} = 3\frac{1}{2}$

Step 4: You cannot leave an improper fraction as the answer. You must change it to a mixed number.

Multiply and reduce answers to lowest terms.

1. $3\frac{1}{5} \times 1\frac{1}{2}$

2. $\frac{3}{8} \times 3\frac{1}{5}$

3. $4\frac{1}{3} \times 2\frac{1}{4}$

4. $4\frac{2}{3} \times 3\frac{3}{4}$

5. $1\frac{1}{2} \times 1\frac{2}{5}$

6. $3\frac{3}{5} \times \frac{5}{6}$

7. $3 \times 6\frac{1}{3}$

8. $6\frac{2}{5} \times 5$

9. $6 \times 1\frac{3}{8}$

10. $\frac{3}{4} \times 2\frac{1}{3}$

11. $1\frac{2}{5} \times 1\frac{1}{4}$

12. $2\frac{1}{2} \times 5\frac{4}{5}$

13. $7\frac{2}{3} \times \frac{3}{4}$

14. $2 \times 3\frac{1}{4}$

15. $2\frac{3}{4} \times 6\frac{2}{3}$

16. $5\frac{3}{5} \times 2\frac{1}{2}$

17. $3\frac{1}{3} \times 3\frac{3}{4}$

18. $5\frac{1}{4} \times 1\frac{2}{3}$

19. $6\frac{2}{5} \times 1\frac{9}{16}$

20. $2\frac{2}{3} \times 2\frac{1}{10}$

21. $3\frac{3}{10} \times 2\frac{2}{3}$

DIVIDING FRACTIONS

EXAMPLE: $1\frac{3}{4} \div 2\frac{5}{8}$

Step 1:   Change the mixed numbers in the problem to improper fractions. $\frac{7}{4} \div \frac{21}{8}$

Step 2:   Invert (turn upside down) the **second** fraction and multiply. $\frac{7}{4} \times \frac{8}{21}$

Step 3:   Cancel where possible and multiply. $\frac{\overset{1}{\cancel{7}}}{\underset{1}{4}} \times \frac{\overset{2}{\cancel{8}}}{\underset{3}{\cancel{21}}} = \frac{2}{3}$

Divide and reduce answers to lowest terms.

1. $1\frac{7}{8} \div 2\frac{1}{4} =$

2. $8 \div 2\frac{2}{3} =$

3. $1\frac{1}{3} \div 1\frac{3}{5} =$

4. $2\frac{1}{4} \div \frac{3}{10} =$

5. $3\frac{5}{6} \div \frac{5}{6} =$

6. $5\frac{1}{3} \div 2\frac{2}{5} =$

7. $16 \div \frac{2}{3} =$

8. $\frac{2}{5} \div 4 =$

9. $8\frac{1}{3} \div 1\frac{1}{4} =$

10. $18 \div 1\frac{1}{5} =$

11. $1\frac{3}{8} \div 3\frac{2}{3} =$

12. $4\frac{3}{4} \div \frac{1}{8} =$

13. $3\frac{3}{8} \div 4\frac{1}{2} =$

14. $2\frac{1}{2} \div 1\frac{2}{3} =$

15. $5\frac{5}{6} \div 1\frac{2}{3} =$

16. $15 \div \frac{5}{2} =$

17. $1\frac{7}{8} \div \frac{5}{6} =$

18. $2\frac{2}{3} \div 3\frac{1}{5} =$

19. $3\frac{1}{3} \div 2\frac{1}{2} =$

20. $4\frac{1}{6} \div \frac{5}{3} =$

21. $12\frac{1}{4} \div 3\frac{1}{2} =$

ORDERING FRACTIONS

If the four fractions given are $\frac{3}{10}$, $\frac{5}{6}$, $\frac{1}{2}$, and $\frac{9}{10}$, the greatest, $\frac{9}{10}$, and least, $\frac{3}{10}$, are easy to find. If the four fractions given are $\frac{3}{5}$, $\frac{5}{6}$, $\frac{1}{2}$, and $\frac{11}{15}$, you probably need to find a common denominator first to figure out which fraction is the greatest or least.

$\frac{3}{5} = \frac{18}{30}$

$\frac{5}{6} = \frac{25}{30}$

$\frac{1}{2} = \frac{15}{30}$

$\frac{11}{15} = \frac{22}{30}$

Ask yourself what number you can divide by 5, 6, 2, and 15 without a remainder.

$5\overline{)\ ?\ }$ $6\overline{)\ ?\ }$ $2\overline{)\ ?\ }$ $15\overline{)\ ?\ }$

The number 30 is the smallest number that can be divided by 5, 6, 2, and 15. Once equivalent fractions with the same denominator are found, you can look at the top number to tell which fraction is greatest, $\frac{5}{6}$ or least, $\frac{1}{2}$.

Circle the greatest fraction, and underline the least fraction in each group below.

1. $\frac{1}{4}$	3. $\frac{2}{3}$	5. $\frac{7}{8}$	7. $\frac{5}{6}$	9. $\frac{5}{8}$
$\frac{3}{8}$	$\frac{1}{6}$	$\frac{5}{12}$	$\frac{1}{2}$	$\frac{2}{3}$
$\frac{1}{12}$	$\frac{4}{9}$	$\frac{3}{4}$	$\frac{3}{4}$	$\frac{5}{16}$
$\frac{5}{6}$	$\frac{1}{2}$	$\frac{5}{8}$	$\frac{7}{12}$	$\frac{1}{2}$

2. $\frac{3}{5}$	4. $\frac{3}{5}$	6. $\frac{5}{12}$	8. $\frac{13}{16}$	10. $\frac{5}{16}$
$\frac{7}{10}$	$\frac{13}{16}$	$\frac{1}{4}$	$\frac{3}{4}$	$\frac{7}{8}$
$\frac{11}{16}$	$\frac{3}{10}$	$\frac{5}{8}$	$\frac{3}{8}$	$\frac{1}{4}$
$\frac{3}{4}$	$\frac{4}{5}$	$\frac{1}{2}$	$\frac{7}{12}$	$\frac{1}{2}$

Find common denominators to determine which fraction is between the two given fractions. Circle the answer.

11. Which fraction is between $\frac{1}{3}$ and $\frac{5}{8}$? $\frac{2}{3}$, $\frac{3}{4}$, $\frac{5}{6}$, or $\frac{3}{8}$

12. Which fraction is between $\frac{1}{2}$ and $\frac{5}{8}$? $\frac{1}{3}$, $\frac{3}{4}$, $\frac{3}{8}$, or $\frac{9}{16}$

13. Which fraction is between $\frac{3}{5}$ and $\frac{7}{8}$? $\frac{1}{4}$, $\frac{3}{4}$, $\frac{5}{12}$, or $\frac{7}{12}$

14. Which fraction is between $\frac{2}{3}$ and $\frac{5}{6}$? $\frac{1}{4}$, $\frac{9}{10}$, $\frac{2}{5}$, or $\frac{4}{5}$

FINDING EQUIVALENT FRACTIONS

REMEMBER: Any fraction that has the same non-zero numerator (top number) and denominator (bottom number) equals 1.

EXAMPLES: $\frac{5}{5} = 1$ $\frac{8}{8} = 1$ $\frac{12}{12} = 1$ $\frac{15}{15} = 1$ $\frac{25}{25} = 1$

Any fraction multiplied by 1 in any fraction form remains equal.

EXAMPLES: $\frac{3}{8} \times \frac{4}{4} = \frac{12}{32}$ so $\frac{3}{8} = \frac{12}{32}$

PROBLEM: Find the missing numerator (top number). $\frac{5}{8} = \frac{\blacksquare}{24}$

Step 1: Ask yourself, "What was 8 multiplied by to get 24?" 3 is the answer.

Step 2: The only way to keep the fraction equal is to multiply the top and bottom number by the same number. The bottom number was multiplied by 3, so multiply the top number by 3. $\frac{5}{8} \times \frac{3}{3} = \frac{15}{24}$

Find the numerators missing from the following equivalent fractions.

1. $\frac{3}{5} = \frac{}{15}$ 7. $\frac{2}{8} = \frac{}{16}$ 13. $\frac{2}{5} = \frac{}{45}$ 19. $\frac{7}{8} = \frac{}{24}$ 25. $\frac{4}{5} = \frac{}{45}$

2. $\frac{3}{8} = \frac{}{32}$ 8. $\frac{5}{6} = \frac{}{18}$ 14. $\frac{3}{12} = \frac{}{24}$ 20. $\frac{5}{8} = \frac{}{24}$ 26. $\frac{3}{8} = \frac{}{40}$

3. $\frac{7}{10} = \frac{}{50}$ 9. $\frac{5}{8} = \frac{}{32}$ 15. $\frac{8}{9} = \frac{}{45}$ 21. $\frac{1}{4} = \frac{}{12}$ 27. $\frac{2}{5} = \frac{}{25}$

4. $\frac{5}{8} = \frac{}{24}$ 10. $\frac{2}{3} = \frac{}{12}$ 16. $\frac{5}{16} = \frac{}{32}$ 22. $\frac{1}{3} = \frac{}{9}$ 28. $\frac{5}{12} = \frac{}{36}$

5. $\frac{3}{4} = \frac{}{16}$ 11. $\frac{1}{4} = \frac{}{12}$ 17. $\frac{4}{5} = \frac{}{20}$ 23. $\frac{2}{5} = \frac{}{35}$ 29. $\frac{3}{4} = \frac{}{28}$

6. $\frac{5}{6} = \frac{}{30}$ 12. $\frac{2}{3} = \frac{}{33}$ 18. $\frac{2}{3} = \frac{}{27}$ 24. $\frac{1}{6} = \frac{}{36}$ 30. $\frac{7}{8} = \frac{}{40}$

ADDING FRACTIONS

EXAMPLE: Add $3\frac{1}{2} + 2\frac{2}{3}$

Step 1: Rewrite the problem vertically, and find a common denominator.

Think: What is the smallest number I can divide 2 and 3 into without a remainder? $2\overline{)?}$ $3\overline{)?}$ 6, of course.

$$3\frac{1}{2} = 3\frac{}{6}$$
$$+\,2\frac{2}{3} = 2\frac{}{6}$$
$$\overline{}$$

Step 2: Find the missing numerators (top numbers) in the same way you did on the previous page.

$$3\frac{1}{2} = 3\frac{3}{6}$$
$$+\,2\frac{2}{3} = 2\frac{4}{6}$$
$$\overline{} = 5\frac{7}{6}$$

Step 3: Add whole numbers and fractions and simplify.

$$= 6\frac{1}{6}$$

Add and simplify the answers.

1. $1\frac{3}{4}$
$+\,2\frac{7}{8}$

5. $8\frac{3}{4}$
$+\,9\frac{2}{3}$

9. $3\frac{1}{6}$
$+\,8\frac{9}{10}$

13. $9\frac{7}{8}$
$+\,3\frac{1}{3}$

17. $4\frac{7}{12}$
$+\,8\frac{2}{3}$

2. $4\frac{1}{5}$
$+\,5\frac{1}{4}$

6. $5\frac{5}{6}$
$+\,7\frac{1}{4}$

10. $6\frac{1}{2}$
$+\,2\frac{4}{5}$

14. $6\frac{1}{6}$
$+\,5\frac{7}{8}$

18. $8\frac{2}{3}$
$+\,9\frac{7}{8}$

3. $6\frac{2}{3}$
$+\,3\frac{2}{5}$

7. $5\frac{2}{3}$
$+\,7\frac{5}{6}$

11. $4\frac{5}{12}$
$+\,8\frac{5}{6}$

15. $7\frac{3}{5}$
$+\,\frac{2}{3}$

19. $7\frac{3}{4}$
$+\,6\frac{1}{8}$

4. $4\frac{3}{5}$
$+\,1\frac{1}{2}$

8. $1\frac{5}{12}$
$+\,6\frac{5}{8}$

12. $9\frac{1}{4}$
$+\,1\frac{5}{12}$

16. $7\frac{4}{5}$
$+\,3\frac{1}{3}$

20. $3\frac{1}{8}$
$+\,9\frac{3}{4}$

SUBTRACTING MIXED NUMBERS FROM WHOLE NUMBERS

EXAMPLE: Subtract $15 - 3\frac{3}{4}$

$$\begin{array}{r} 15 \\ -\ 3\frac{3}{4} \\ \hline \end{array}$$

Step 1: Rewrite the problem vertically.

Step 2: You cannot subtract three-fourths from nothing. You must borrow 1 from 15. You will need to put the 1 in fraction form. If you use $\frac{4}{4}$ ($\frac{4}{4} = 1$), you will be ready to subtract.

$$\begin{array}{r} \overset{14}{\cancel{15}}\ \frac{4}{4} \\ -\ 3\ \frac{3}{4} \\ \hline 11\ \frac{1}{4} \end{array}$$

Subtract.

1. $\begin{array}{r} 21 \\ -\ 6\frac{5}{7} \\ \hline \end{array}$
5. $\begin{array}{r} 14 \\ -\ 9\frac{7}{8} \\ \hline \end{array}$
9. $\begin{array}{r} 8 \\ -\ 7\frac{1}{2} \\ \hline \end{array}$
13. $\begin{array}{r} 11 \\ -\ 7\frac{5}{6} \\ \hline \end{array}$
17. $\begin{array}{r} 6 \\ -\ 4\frac{2}{3} \\ \hline \end{array}$
21. $\begin{array}{r} 15 \\ -\ 13\frac{4}{5} \\ \hline \end{array}$
25. $\begin{array}{r} 12 \\ -\ 8\frac{3}{10} \\ \hline \end{array}$

2. $\begin{array}{r} 7 \\ -\ 4\frac{1}{4} \\ \hline \end{array}$
6. $\begin{array}{r} 16 \\ -\ 12\frac{5}{8} \\ \hline \end{array}$
10. $\begin{array}{r} 22 \\ -\ 14\frac{1}{6} \\ \hline \end{array}$
14. $\begin{array}{r} 31 \\ -\ 26\frac{2}{5} \\ \hline \end{array}$
18. $\begin{array}{r} 10 \\ -\ 7\frac{1}{4} \\ \hline \end{array}$
22. $\begin{array}{r} 5 \\ -\ 1\frac{3}{8} \\ \hline \end{array}$
26. $\begin{array}{r} 13 \\ -\ 11\frac{4}{5} \\ \hline \end{array}$

3. $\begin{array}{r} 25 \\ -\ 14\frac{7}{8} \\ \hline \end{array}$
7. $\begin{array}{r} 16 \\ -\ 7\frac{3}{10} \\ \hline \end{array}$
11. $\begin{array}{r} 7 \\ -\ 4\frac{2}{3} \\ \hline \end{array}$
15. $\begin{array}{r} 23 \\ -\ 15\frac{1}{5} \\ \hline \end{array}$
19. $\begin{array}{r} 34 \\ -\ 26\frac{3}{8} \\ \hline \end{array}$
23. $\begin{array}{r} 51 \\ -\ 33\frac{9}{10} \\ \hline \end{array}$
27. $\begin{array}{r} 30 \\ -\ 18\frac{5}{6} \\ \hline \end{array}$

4. $\begin{array}{r} 16 \\ -\ 14\frac{1}{5} \\ \hline \end{array}$
8. $\begin{array}{r} 25 \\ -\ 17\frac{5}{8} \\ \hline \end{array}$
12. $\begin{array}{r} 33 \\ -\ 25\frac{7}{12} \\ \hline \end{array}$
16. $\begin{array}{r} 19 \\ -\ 7\frac{4}{5} \\ \hline \end{array}$
20. $\begin{array}{r} 47 \\ -\ 38\frac{4}{7} \\ \hline \end{array}$
24. $\begin{array}{r} 18 \\ -\ 5\frac{3}{8} \\ \hline \end{array}$
28. $\begin{array}{r} 34 \\ -\ 18\frac{7}{12} \\ \hline \end{array}$

SUBTRACTING MIXED NUMBERS WITH BORROWING

EXAMPLE: Subtract $7\frac{1}{4} - 5\frac{5}{6}$

Step 1: Rewrite the problem, and find a common denominator.

$$7\frac{1}{4} = \frac{3}{12}$$
$$-5\frac{5}{6} = \frac{10}{12}$$

Step 2: You cannot subtract 10 from 3. You must borrow 1 from the 7. The 1 will be in the fraction form $\frac{12}{12}$ which you must add to the $\frac{3}{12}$ you already have, making $\frac{15}{12}$.

$$\overset{6}{\cancel{7}}\frac{1}{4} = \frac{\overset{15}{\cancel{3}}}{12}$$
$$-5\frac{5}{6} = \frac{10}{12}$$
$$\overline{1\frac{5}{12}}$$

Step 3: Add whole numbers and fractions and simplify.

Subtract and simplify.

1. $9\frac{2}{3}$
   $-4\frac{3}{4}$

4. $3\frac{1}{6}$
   $-2\frac{3}{4}$

7. $7\frac{1}{4}$
   $-5\frac{4}{9}$

10. $5\frac{1}{2}$
    $-3\frac{2}{3}$

13. $2\frac{3}{8}$
    $-1\frac{5}{6}$

16. $4\frac{1}{5}$
    $-2\frac{7}{8}$

2. $6\frac{5}{8}$
   $-3\frac{5}{6}$

5. $10\frac{7}{10}$
   $-5\frac{4}{5}$

8. $1\frac{1}{6}$
   $-\frac{7}{8}$

11. $3\frac{1}{9}$
    $-2\frac{2}{3}$

14. $4\frac{3}{8}$
    $-3\frac{2}{3}$

17. $2\frac{1}{5}$
    $-1\frac{3}{4}$

3. $4\frac{5}{8}$
   $-2\frac{2}{3}$

6. $9\frac{1}{3}$
   $-3\frac{1}{2}$

9. $2\frac{1}{6}$
   $-1\frac{4}{5}$

12. $4\frac{2}{3}$
    $-2\frac{3}{4}$

15. $7\frac{3}{8}$
    $-5\frac{2}{3}$

18. $6\frac{1}{3}$
    $-5\frac{7}{8}$

FRACTION REVIEW

Change to mixed numbers or whole numbers.

1. $\frac{11}{2} =$ 3. $\frac{9}{4} =$ 5. $\frac{16}{4} =$ 7. $\frac{13}{4} =$ 9. $\frac{30}{10} =$ 11. $\frac{10}{4} =$

2. $\frac{8}{3} =$ 4. $\frac{20}{9} =$ 6. $\frac{21}{5} =$ 8. $\frac{42}{7} =$ 10. $\frac{18}{7} =$ 12. $\frac{11}{3} =$

Change to an improper fraction.

13. $2\frac{1}{8} =$ 15. $4\frac{2}{3} =$ 17. $9\frac{2}{5} =$ 19. $6\frac{2}{7} =$ 21. $2\frac{5}{8} =$ 23. $8\frac{1}{9} =$

14. $8\frac{1}{4} =$ 16. $9\frac{1}{3} =$ 18. $6\frac{3}{4} =$ 20. $2\frac{3}{4} =$ 22. $9\frac{1}{2} =$ 24. $7\frac{2}{7} =$

Reduce to lowest terms.

25. $\frac{2}{8} =$ 27. $\frac{5}{15} =$ 29. $\frac{10}{12} =$ 31. $\frac{6}{18} =$ 33. $\frac{4}{8} =$ 35. $\frac{6}{14} =$

26. $\frac{10}{15} =$ 28. $\frac{6}{24} =$ 30. $\frac{16}{20} =$ 32. $\frac{8}{24} =$ 34. $\frac{8}{14} =$ 36. $\frac{7}{28} =$

Perform the following operations. Reduce each answer to lowest terms.

37. $\begin{array}{r} 5\frac{1}{2} \\ + 3\frac{5}{8} \\ \hline \end{array}$ 38. $\begin{array}{r} 7\frac{3}{8} \\ + 4\frac{1}{2} \\ \hline \end{array}$ 39. $\begin{array}{r} 9\frac{3}{4} \\ + 4\frac{1}{2} \\ \hline \end{array}$ 40. $\begin{array}{r} 2\frac{2}{3} \\ + 2\frac{1}{4} \\ \hline \end{array}$ 41. $\begin{array}{r} 4\frac{5}{6} \\ + 6\frac{1}{4} \\ \hline \end{array}$

42. $\begin{array}{r} 7 \\ - 3\frac{1}{4} \\ \hline \end{array}$ 43. $\begin{array}{r} 6\frac{1}{4} \\ - 2\frac{3}{5} \\ \hline \end{array}$ 44. $\begin{array}{r} 9\frac{1}{8} \\ - 7\frac{3}{4} \\ \hline \end{array}$ 45. $\begin{array}{r} 5\frac{1}{4} \\ - 2\frac{2}{3} \\ \hline \end{array}$ 46. $\begin{array}{r} 8\frac{1}{8} \\ - 2\frac{1}{10} \\ \hline \end{array}$

47. $\frac{3}{4} \times 100$ 48. $\frac{2}{5} \times 80$ 49. $\frac{3}{4} \times \frac{7}{9}$ 50. $\frac{1}{3} \times 16\frac{1}{2}$

51. $10\frac{1}{2} \times 4\frac{2}{3}$ 52. $4\frac{1}{6} \times 2\frac{4}{5}$ 53. $9 \times 7\frac{2}{3}$ 54. $2 \times 1\frac{7}{8}$

55. $28 \div 2\frac{2}{3}$ 56. $1\frac{1}{8} \div 2$ 57. $1\frac{1}{4} \div 2\frac{2}{9}$

58. $15 \div 1\frac{1}{3}$ 59. $8\frac{1}{4} \div \frac{1}{2}$ 60. $20 \div 2\frac{2}{5}$

COMPARING THE RELATIVE MAGNITUDE OF FRACTIONS

Comparing the relative magnitude of fractions using the greater than (>), less than (<), and equal to (=) signs.

EXAMPLE 1: Compare $\frac{3}{4}$ and $\frac{5}{8}$.

Step 1: Find the lowest common denominator. The lowest common denominator is 8.

Step 2: Change fourths to eighths by multiplying three fourths by two halves, $\frac{2 \times 3}{2 \times 4} = \frac{6}{8}$.

Step 3: $\frac{6}{8} > \frac{5}{8}$

EXAMPLE 2: Compare the mixed numbers $1\frac{3}{5}$ and $1\frac{2}{3}$.

Step 1: Change the mixed numbers to improper fractions (explained in a previous lesson).

$$1\frac{3}{5} = \frac{8}{5} \quad \text{and} \quad 1\frac{2}{3} = \frac{5}{3}$$

Step 2: Find the lowest common denominator for the improper fractions. The lowest common denominator is 15.

Step 3: Change fifths to fifteenths and thirds to fifteenths, $\frac{3 \times 8}{3 \times 5} = \frac{24}{15}$ and $\frac{5 \times 5}{5 \times 3} = \frac{25}{15}$.

Step 4: $\frac{24}{15} < \frac{25}{15}$ therefore $1\frac{3}{5} < 1\frac{2}{3}$.

Fill in the box with the correct sign.

1. $\frac{7}{9} \ \square \ \frac{7}{8}$

2. $\frac{6}{7} \ \square \ \frac{5}{6}$

3. $\frac{4}{6} \ \square \ \frac{5}{7}$

4. $\frac{3}{10} \ \square \ \frac{4}{13}$

5. $\frac{5}{8} \ \square \ \frac{4}{11}$

6. $\frac{5}{8} \ \square \ \frac{4}{7}$

7. $\frac{9}{10} \ \square \ \frac{8}{13}$

8. $\frac{2}{13} \ \square \ \frac{1}{10}$

9. $\frac{4}{9} \ \square \ \frac{3}{5}$

10. $\frac{2}{6} \ \square \ \frac{4}{5}$

11. $\frac{7}{12} \ \square \ \frac{6}{11}$

12. $\frac{3}{11} \ \square \ \frac{5}{12}$

DEDUCTIONS - FRACTION OFF

Sometimes sale prices are advertised as $\frac{1}{4}$ off or $\frac{1}{3}$ off. To find out how much you will save, just multiply the original price by the fraction off.

EXAMPLE: CD players are on sale for $\frac{1}{3}$ off. How much can you save on a $240 CD player?

$$\frac{1}{\overset{}{\underset{1}{\cancel{3}}}} \times \frac{\overset{80}{\cancel{240}}}{1} = 80. \text{ You can save \$80.00}$$

Find the amount of savings in the problems below.

J.P. Nichols is having a liquidation sale on all furniture. Sale prices are $\frac{1}{2}$ off the regular price. How much can you save on the following furniture items?

Item	Regular Price	Savings
1. Couch	$850	_____
2. Loveseat	$624	_____
3. Recliner	$457	_____
4. Dining Room Set	$1,352	_____
5. Bedroom Set	$2,648	_____

$\frac{1}{2}$ off all items in the store

Buy Rite Computer Store is having a $\frac{1}{3}$ off sale on selected computer items in the store. How much can you save on the following items?

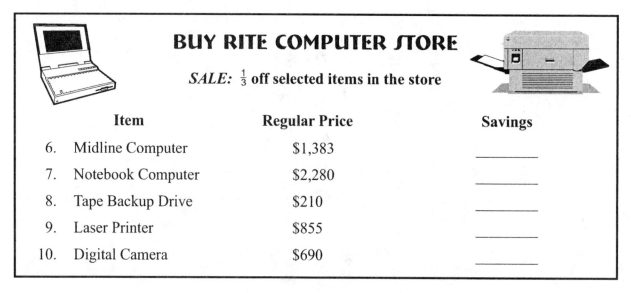

BUY RITE COMPUTER STORE

SALE: $\frac{1}{3}$ off selected items in the store

Item	Regular Price	Savings
6. Midline Computer	$1,383	_____
7. Notebook Computer	$2,280	_____
8. Tape Backup Drive	$210	_____
9. Laser Printer	$855	_____
10. Digital Camera	$690	_____

Martin's Department Store is having its annual $\frac{1}{3}$ off all bedroom apparel sale. How much can you save on the following items?

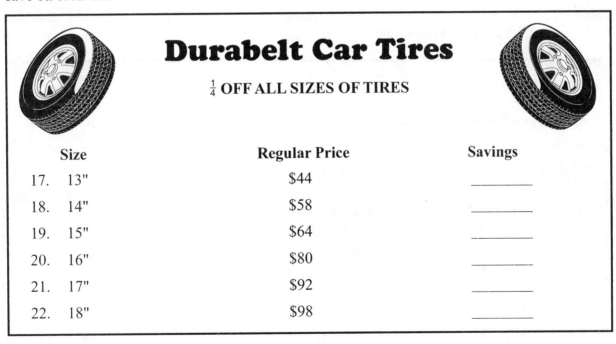

Martin's Department Store

$\frac{1}{3}$ off all "Country Elegant" comforters and accessories

	Length	Regular Price	Savings
11.	Twin	$87	_____
12.	Full	$114	_____
13.	Queen	$141	_____
14.	King	$177	_____
15.	Accent Pillows	$24	_____
16.	Draperies	$78	_____

Durabelt Car Tires is placing its tires on sale for $\frac{1}{4}$ off the regular price. Find the amount you can save on each tire.

Durabelt Car Tires

$\frac{1}{4}$ OFF ALL SIZES OF TIRES

	Size	Regular Price	Savings
17.	13"	$44	_____
18.	14"	$58	_____
19.	15"	$64	_____
20.	16"	$80	_____
21.	17"	$92	_____
22.	18"	$98	_____

FINDING A FRACTION OF A TOTAL

Mathematicians use the word "of" in word problems to indicate that you need to multiply to find the answer.

EXAMPLE 1:   Two-thirds of male high school seniors will be taller than their fathers by the time they graduate. In a sample of 400 male seniors, about how many will be taller than their fathers on graduation day?

Solution: Multiply the fraction by the total. $\dfrac{2}{3} \times \dfrac{400}{1} = \dfrac{800}{3} = 267$

About 267 out of 400 male seniors will be taller than their father.

EXAMPLE 2:

Favorite Cake Flavors

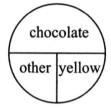

In a lunchroom survey of 360 students, about how many students preferred yellow cake?

Step 1:   Estimate the fraction of the circle that shows the amount of yellow cake. It looks like about $\frac{1}{4}$ of the circle.

Step 2:   Multiply $\frac{1}{4} \times 360$, the total number of students surveyed. About 90 students prefer yellow cake.

Solve the following problems.

1.   This year $\frac{2}{3}$ of the seniors went to the prom. Out of a class of 438 seniors, how many went to the prom?

2.   The North End Diner surveyed its customers to see which flavor of ice cream they preferred. Out of 512 customers, how many preferred chocolate chip?

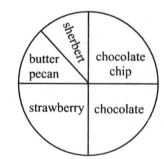

3.   It has rained 5 out of 6 days for the last month! If there were 30 days in the last month, how many days did it rain?

4. Ryan worked 5 hours on his homework Tuesday. Two-thirds of the time was spent on algebra word problems. How much time did he spend on algebra?

5. Beth earned $20 babysitting. She spent $\frac{2}{5}$ of it on a paperback book. How much did she spend on the book?

6. Three-fourths of the graduating class at Lakewood High School plans on going to college. There are 680 graduating seniors. How many are planning on going to college?

7. At West End High School, $\frac{1}{5}$ of the students in the band are also in the choir. There are 205 band members. How many are also in the choir?

8. Allen bought a $3\frac{1}{4}$ pound roast to cook for dinner. How much did the roast cost at $2.00 per pound?

9. What is the cost of $4\frac{1}{3}$ yards of fabric if it sells for $15 per yard?

10. Grandma made $4\frac{1}{2}$ dozen cookies. Her grandsons ate $\frac{2}{3}$ of them. How many cookies did they eat?

11. Randy needed to make a ditch $20\frac{1}{2}$ feet long. The first day he finished $\frac{2}{3}$ of it. How many feet did he finish digging?

12. The band members brought in 300 cupcakes to sell for a fund-raiser. Three-fifths of them were chocolate. How many chocolate cupcakes were there?

FRACTION WORD PROBLEMS

1. Sara Beth bought $\frac{1}{4}$-pound hot dogs for her class picnic. She bought 30 hot dogs. How many pounds of hot dogs did she buy?

2. When Jonathan took a trip, he wanted to check his mileage. At the start of his trip, his gas tank was full. He bought $10\frac{1}{4}$, $12\frac{1}{8}$, $11\frac{3}{4}$ gallons and then $5\frac{3}{8}$ gallons at the end of his trip to fill up the tank again. How much gas did his trip take?

3. Jodene bought a 10-pound bag of flour. She used $3\frac{3}{4}$ pounds to make bread. How much flour did she have left?

4. Cindy needed a total of $5\frac{1}{4}$ yards of trim for cushions she was making. She already had $3\frac{3}{4}$ yards. How much more did she need to buy?

5. Mary bought a 10-ounce bottle of cough syrup. If each dose is $\frac{1}{4}$ ounce, how many doses are in the bottle?

6. It took $7\frac{1}{2}$ hours to go 390 miles on our vacation. What was our average speed in miles per hour?

7. Stella bought a 100-pound sack of flour to make bread for the bakery. If $\frac{7}{8}$ of a pound of flour is needed for each loaf, how many loaves can she make? What fraction of a pound of flour is left over?

8. Seth took a 30-inch piece of wire and cut it into $1\frac{1}{4}$ inch pieces. How many pieces did he get?

9. Keisha wanted to bake 10 cakes for a fund-raiser. She needed $1\frac{3}{4}$ cups of sugar for each cake. How many cups of sugar did she need in all?

10. Mark had $3\frac{1}{2}$ quarts of oil in a can. He put $1\frac{3}{4}$ quarts in his car. How much oil did he have left?

60

11. Sal works for a movie theater and sells candy by the pound. Her first customer bought $1\frac{1}{3}$ pounds of candy, the second bought $\frac{3}{4}$ of a pound, and the third bought $\frac{1}{2}$ of a pound. How many pounds did she sell to the first three customers?

12. Beth has a bread machine that makes a loaf of bread that weighs $1\frac{1}{2}$ pounds. If she makes a loaf of bread for each of her three sisters, how many pounds of bread will she make?

13. A farmer hauled in 120 bales of hay. Each of his cows ate $1\frac{1}{4}$ bales. How many cows did the farmer feed?

14. John was competing in a 1,000-meter race. He had to pull out of the race after running $\frac{3}{4}$ of it. How many meters did he run?

15. Tad needs to measure where the free-throw line should be in front of his basketball goal. He knows his feet are $1\frac{1}{8}$ feet long, and the free-throw line should be 15 feet from the backboard. How many toe-to-toe steps does Tad need to take to mark off 15 feet?

16. A chemical plant takes in $5\frac{1}{2}$ million gallons of water from a local river and discharges $3\frac{2}{3}$ million directly back into the river. How much water does not go directly back into the river?

17. In January, Jeff filled his car with $11\frac{1}{2}$ gallons of gas the first week, $13\frac{1}{3}$ gallons the second week, $12\frac{1}{4}$ gallons the third week, and $10\frac{1}{6}$ gallons the fourth week. How many gallons of gas did he buy in January?

18. Martin makes sandwiches for his family. He has $11\frac{1}{4}$ ounces of sandwich meat. If he divides the meat equally to make $4\frac{1}{2}$ sandwiches, how much meat will each sandwich have?

19. The company water cooler started with $4\frac{1}{3}$ gallons of water. Employees drank $3\frac{3}{4}$ gallons. How many gallons were left in the cooler?

20. Rita bought $\frac{1}{4}$ pound hamburger patties for her family reunion picnic. She bought 50 patties. How many pounds of hamburgers did she buy?

61

CHAPTER 3 REVIEW

1. Simplify: $\dfrac{12}{5}$ _____

2. Reduce: $\dfrac{25}{3}$ _____

3. Change to an improper fraction: $2\dfrac{7}{8}$ _____

4. Change to an improper fraction: 5 _____

5. Add: $\dfrac{3}{8} + \dfrac{7}{8}$ _____

6. Add: $10\dfrac{3}{4} + 2\dfrac{2}{3}$ _____

7. Subtract: $12 - 6\dfrac{5}{8}$ _____

8. Subtract: $5\dfrac{1}{8} - 2\dfrac{5}{6}$ _____

9. Multiply: $8 \times 4\dfrac{3}{4}$ _____

10. Multiply: $2\dfrac{2}{3} \times 3\dfrac{3}{4}$ _____

11. Divide: $7 \div 1\dfrac{3}{4}$ _____

12. Divide: $1\dfrac{2}{5} \div 2\dfrac{1}{10}$ _____

13. Tami wants to buy a sweater that is on sale for $\frac{1}{4}$ off the regular $56.00 price. How much will she save?

_____

14. Whitlow's has a $\frac{3}{4}$ off the regular price clearance sale on dress shirts. How much can you save if you buy a shirt regularly priced at $32.00?

_____

15. Mrs. Tate brought $5\frac{1}{2}$ pounds of candy to divide among her 22 students. If the candy was divided equally, how much did each student's portion weigh?

_____

16. Dorothy used $1\frac{1}{4}$ yards of material to recover one dining room chair. How much material would she need to recover all eight chairs?

_____

17. The square tiles in Mr. Cooke's math classroom measure $2\frac{1}{4}$ feet across. The classroom is $5\frac{1}{3}$ tiles wide. How many feet wide is Mr. Cooke's classroom?

_____

18. Yvonne biked for $1\frac{1}{4}$ hours and traveled $12\frac{1}{2}$ miles. How many miles per hour did she average?

_____

19. Jason used $\frac{1}{3}$ of a pound of hamburger to make each hamburger patty he was grilling. He needs to make 42 hamburgers. How many pounds of meat will he need?

_____

24.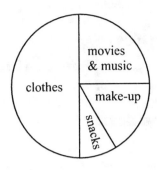

April gets $150 per month allowance from her parents. The graph above shows how she spends it. About how much does she spend on movies and music?

_____

20. **Johnson Family Budget**

The diagram above shows where the money goes in the Johnson family. If the family brings home a total of $4,212 each month, about how much of the money goes into savings?

_____

25. Joni walked $1\frac{1}{2}$ miles on Monday, $2\frac{1}{4}$ miles on Wednesday, and $1\frac{2}{3}$ miles on Friday. How many miles did she walk total?

_____

21. Erica is $61\frac{3}{4}$ inches tall. Her older brother is now $67\frac{1}{4}$ inches tall. How much taller is her brother?

_____

22. How many student lunches were sold if $\frac{2}{3}$ of the 1,263 students bought their lunch?

_____

23. How many $3\frac{1}{2}$-foot lengths of rope can be cut from 56 feet of rope?

_____

RATIOS, PROBABILITY, PROPORTIONS, AND SCALE DRAWINGS

RATIO PROBLEMS

In some word problems, you may be asked to express answers as a **ratio**. Ratios can look like fractions, they can be written with a colon, or they can be written in word form with "to" between the numbers. Numbers must be written in the order they are requested. In the following problem, 8 cups of sugar are mentioned before 6 cups of strawberries. But in the question part of the problem, you are asked for the ratio of STRAWBERRIES to SUGAR. The amount of strawberries IS THE FIRST WORD MENTIONED, so it must be the **top** number of the fraction. The amount of sugar, THE SECOND WORD MENTIONED, must be the **bottom** number of the fraction.

EXAMPLE: The recipe for jam requires 8 cups of sugar for every 6 cups of strawberries. What is the ratio of strawberries to sugar in this recipe?

First number requested $\dfrac{6}{8}$ **cups strawberries**
Second number requested   **cups sugar**

Answers may be reduced to lowest terms. $\dfrac{6}{8} = \dfrac{3}{4}$

This ratio is also correctly expressed as 3:4 or 3 to 4.

Practice writing ratios for the following word problems, and reduce to lowest terms. DO NOT CHANGE ANSWERS TO MIXED NUMBERS. Ratios should be left in fraction form.

1. Out of the 248 seniors, 112 are boys. What is the ratio of boys to the total number of seniors?

   _____

2. It takes 7 cups of flour to make 2 loaves of bread. What is the ratio of cups of flour to loaves of bread?

   _____

3. A skyscraper that stands 620 feet tall casts a shadow that is 125 feet long. What is the ratio of the shadow to the height of the skyscraper?

   _____

4. The newborn weighs 8 pounds and is 22 inches long. What is the ratio of weight to length?

   _____

5. Jack paid $6.00 for 10 pounds of apples. What is the ratio of the price of apples to the pounds of apples?

   _____

6. Twenty boxes of paper weigh 520 pounds. What is the ratio of boxes to pounds?

   _____

PROBABILITY

Probability is the chance that something will happen. Probability is most often expressed as a fraction but can also be written out in words.

EXAMPLE 1: Billy had 3 red marbles, 5 white marbles, and 4 blue marbles on the floor. His cat came along and batted one marble under the chair. What is the **probability** it was a red marble?

Step 1: The number of red marbles will be the top number of the fraction. $\longrightarrow$ $\dfrac{3}{12}$

Step 2: The total number of marbles is the bottom number of the fraction. $\longrightarrow$

The answer may be expressed in lowest terms. $\dfrac{3}{12} = \dfrac{1}{4}$

Expressed in words, the answer is **one out of four**.

EXAMPLE 2: Determine the probability that the pointer will stop on a shaded wedge or the number 1.

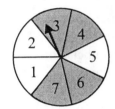

Step 1: Count the number of possible wedges that the spinner can stop on to satisfy the above problem. There are 5 wedges that satisfy it (4 shaded wedges and one number 1). The top number of the fraction is 5.

Step 2: Count the total number of wedges, 7. The bottom number of the fraction is 7.

The answer is $\dfrac{5}{7}$ or **five out of seven**.

EXAMPLE 3: Refer to the spinner above. If the pointer stops on the number 7, what is the probability that it will **not** stop on 7 on the next spin?

Ignore the information that the pointer stopped on the number 7 on the previous spin. The probability of the next spin does not depend on the outcome of the previous spin. Simply find the probability that the spinner will **not** stop on 7.

In other words, the question is asking: What is the probability the spinner will land on one of the other 6 wedges?

Therefore, the probability is $\dfrac{6}{7}$ or **six out of seven**.

Find the probability of the following problems.

1. A computer chose a random number between 1 and 50. What is the probability of you guessing the same number that the computer chose?

———

2. There are 24 candy-coated chocolate pieces in a bag. Eight have defects in the coating that can be seen only with close inspection. What is the probability of pulling out a defective piece without looking?

———

3. Seven sisters have to choose which day each will wash the dishes. They put equal-size pieces of paper, each labeled with a day of the week, in a hat. What is the probability that the first sister who draws will choose a weekend day?

———

4. For his garden, Clay has a mixture of 12 white corn seeds, 24 yellow corn seeds, and 16 bi-color corn seeds. If he reaches for a seed without looking, what is the probability that Clay will plant a bi-color corn seed first?

———

5. Mom just got a new department store credit card in the mail. What is the probability that the last digit is an odd number?

———

6. Alex has a paper bag of cookies that includes 8 chocolate chip, 4 peanut butter, 6 butterscotch chip, and 12 ginger. Without looking, his friend John reaches in the bag for a cookie. What is the probability that the cookie is peanut butter?

———

7. An umpire at a Little League baseball game has 14 balls in his pockets. Five of the balls are brand A, 6 are brand B, and 3 are brand C. What is the probability that the next ball he throws to the pitcher is a brand C ball?

———

8. What is the probability that the spinner arrow will land on an even number?

———

9. The spinner in the problem above stopped on a shaded wedge on the first spin and stopped on the number 2 on the second spin. What is the probability that it will not stop on a shaded wedge or on the 2 on the third spin?

———

10. A company is offering 1 grand prize, 3 second-place prizes, and 25 third-place prizes based on a random drawing of contest entries. If you entered one of the total 500 entries, what is the probability you will win a third-place prize?

———

11. In the contest problem above, what is the probability that you will win the grand prize or a second-place prize?

———

12. A box of a dozen doughnuts has 3 lemon cream-filled, 5 chocolate cream-filled, and 4 vanilla cream-filled. If the donuts look identical, what is the probability of picking a lemon cream-filled?

———

SOLVING PROPORTIONS

Two **ratios (fractions)** that are **equal** to each other are called **proportions**. For example, $\frac{1}{4} = \frac{2}{8}$. Read the following example to see how to find a number missing from a proportion.

EXAMPLE: $\frac{5}{15} = \frac{8}{x}$

Step 1: To find x, you first multiply the two numbers that are diagonal to each other. $\quad$ **15 × 8 = 120** $\qquad$ $\frac{5}{\boxed{15}} = \frac{\boxed{8}}{x}$

Step 2: Then divide the product (120) by the other number in the proportion (5). **120 ÷ 5 = 24**

$$\text{Therefore, } \frac{5}{15} = \frac{8}{24} \quad x = 24$$

Practice finding the number missing from the following proportions. First, multiply the two numbers that are diagonal from each other. Then divide by the other number.

1. $\frac{2}{5} = \frac{6}{x}$

2. $\frac{9}{3} = \frac{x}{5}$

3. $\frac{x}{12} = \frac{3}{4}$

4. $\frac{7}{x} = \frac{3}{9}$

5. $\frac{12}{x} = \frac{2}{5}$

6. $\frac{12}{x} = \frac{4.}{3}$

7. $\frac{27}{3} = \frac{x}{2}$

8. $\frac{1}{x} = \frac{3}{12}$

9. $\frac{15}{2} = \frac{x}{4}$

10. $\frac{7}{14} = \frac{x}{6}$

11. $\frac{5}{6} = \frac{10}{x}$

12. $\frac{4}{x} = \frac{3}{6}$

13. $\frac{x}{5} = \frac{9}{15}$

14. $\frac{9}{18} = \frac{x}{2}$

15. $\frac{5}{7} = \frac{35}{x}$

16. $\frac{x}{2} = \frac{8}{4}$

17. $\frac{15}{20} = \frac{x}{8}$

18. $\frac{x}{40} = \frac{5}{100}$

19. $\frac{4}{7} = \frac{x}{28}$

20. $\frac{7}{6} = \frac{42}{x}$

21. $\frac{x}{8} = \frac{1}{4}$

RATIO AND PROPORTION WORD PROBLEMS

You can use ratios and proportions to solve problems.

EXAMPLE: A stick one meter long is held perpendicular to the ground and casts a shadow 0.4 meters long. At the same time, an electrical tower casts a shadow 112 meters long. Use ratio and proportion to find the height of the tower.

Step 1: Set up a proportion using the numbers in the problem. Put the shadow lengths on one side of the equation and put the heights on the other side. The 1 meter height is paired with the 0.4 meter length, so let them both be top numbers. Let the unknown height be x.

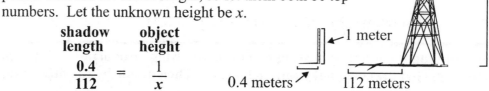

$$\frac{\overset{\textbf{shadow length}}{0.4}}{112} = \frac{\overset{\textbf{object height}}{1}}{x}$$

0.4 meters   112 meters

Step 2: Solve the proportion as you did on the previous page. $112 \times 1 = 112$
$112 \div 0.4 = 280$ **Answer:** The tower height is 280 meters.

Use ratio and proportion to solve the following problems.

1. Rudolph can mow a lawn that measures 1,000 square feet in 2 hours. At that rate, how long would it take him to mow a lawn 3,500 square feet?

   _____

2. Faye wants to know how tall her school building is. On a sunny day, she measures the shadow of the building to be 6 feet. At the same time she measures the shadow cast by a 5-foot statue to be 2 feet. How tall is her school building?

   _____

3. Out of every 5 students surveyed, 2 listen to country music. At that rate, how many students in a school of 800 listen to country music?

   _____

4. Bailey, a Labrador Retriever, had a litter of 8 puppies. Four were black. At that rate, how many would be black in a litter of 10 puppies?

   _____

5. According to the instructions on a bag of fertilizer, 5 pounds of fertilizer are needed for every 100 square feet of lawn. How many square feet will a 25 pound bag cover?

   _____

6. A race car can travel 2 laps in 5 minutes. How long will it take the race car to complete 100 laps at that rate?

   _____

7. If it takes 7 cups of flour to make 4 loaves of bread, how many loaves of bread can you make from 35 cups of flour?

   _____

8. If 3 pounds of jelly beans cost $6.30, how much would 2 pounds cost?

   _____

9. For the first 4 home football games, the concession stand sold 600 hotdogs. If that ratio stays constant, how many hotdogs will sell for all 10 home games?

   _____

MAPS AND SCALE DRAWINGS

EXAMPLE 1: On a map drawn to scale, 5 cm represents 30 kilometers. A line segment connecting two cities is 7 cm long. What distance does this line segment represent?

Step 1: Set up a proportion using the numbers in the problem. Keep centimeters on one side of the equation and kilometers on the other. The 5 cm is paired with the 30 kilometers, so let them both be top numbers. Let the unknown distance be x.

$$\overset{\textbf{cm}}{\frac{5}{7}} = \overset{\textbf{km}}{\frac{30}{x}}$$

Step 2: Solve the proportion as you have previously. $7 \times 30 = 210$

$210 \div 5 = 42$ **Answer:** 7 cm represents 42 km.

Sometimes the answer to a scale drawing problem will be a fraction or mixed number.

EXAMPLE 2: On a scale drawing, 2 inches represent 30 feet. How many inches long is a line segment that represents 5 feet?

Step 1: Set up the proportion as you did above.

$$\overset{\textbf{inches}}{\frac{2}{x}} = \overset{\textbf{feet}}{\frac{30}{5}}$$

Step 2: **First multiply the two numbers that are diagonal from each other. Then divide by the other number.**

$2 \times 5 = 10$ $10 \div 30$ is less than 1, so express the answer as a fraction and reduce.

$$10 \div 30 = \frac{10}{30} = \frac{1}{3} \text{ inch}$$

Set up proportions for each of the following problems and solve.

1. If 2 inches represents 50 miles on a scale drawing, how long would a line segment be that represents 25 miles?

2. On a scale drawing, 2 cm represents 15 km. A line segment on the drawing is 3 cm long. What distance does this line segment represent?

3. On a map drawn to scale, 5 cm represents 250 km. How many kilometers are represented by a line 6 cm long?

4. If 2 inches represents 80 miles on a scale drawing, how long would a line segment be that represents 280 miles?

5. On a map drawn to scale, 5 cm represents 200 km. How long would a line segment be that represents 260 km?

6. On a scale drawing of a house plan, one inch represents 5 feet. How many feet wide is the bathroom if the width on the drawing is 3 inches?

USING A SCALE TO FIND DISTANCES

By using a **map scale**, you can determine the distance between two places in the real world. The **map scale** shows distances in both miles and kilometers. You will need your ruler to do these exercises. On the scale below, you will notice that 1 inch = 800 miles. To find the distance between Calgary and Ottawa, measure with a ruler between the two cities. You will find it measures about $2\frac{1}{2}$ inches. From the scale, you know 1 inch = 800 miles. Use multiplication to find the distance in miles. $2.5 \times 800 = 2,000$. The cities are about 2,000 miles apart.

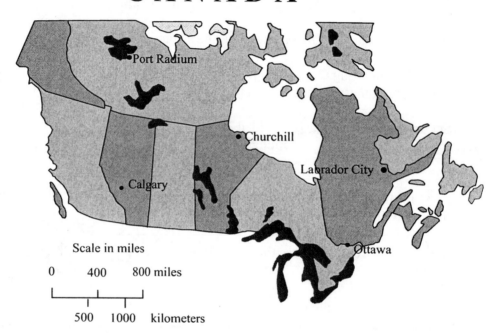

Find these distances in miles.

1. Calgary to Churchill _____

2. Churchill to Ottawa _____

3. Port Radium to Churchill _____

4. Port Radium to Ottawa _____

5. Labrador City to Ottawa _____

6. Calgary to Labrador City _____

Find these distances in kilometers.

7. Churchill to Labrador City _____

8. Ottawa to Port Radium _____

9. Port Radium to Calgary _____

10. Churchill to Ottawa _____

11. Calgary to Churchill _____

12. Calgary to Ottawa _____

USING A SCALE ON A BLUEPRINT

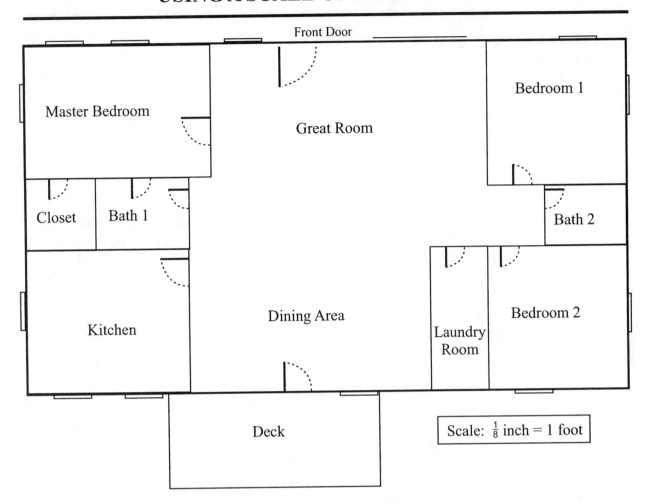

Use a ruler to find the measurements of the rooms on the blueprint above. Convert to feet using the scale. The first problem is done for you.

	long wall		short wall	
	ruler measurement	room measurement	ruler measurement	room measurement
1. Kitchen	$1\frac{3}{4}$ inch	14 feet	$1\frac{1}{2}$ inch	12 feet
2. Deck				
3. Closet				
4. Bedroom 1				
5. Bedroom 2				
6. Master Bedroom				
7. Bath 1				
8. Bath 2				

CHAPTER 4 REVIEW

1. Out of 100 coins, 45 are in mint condition. What is the ratio of mint condition coins to the total number of coins?

   _____

2. What is the probability that the spinner will stop on a shaded wedge or an odd number?

   _____

3. Twenty out of the total 235 seniors graduated with honors. What is the ratio of seniors graduating with honors to the total number of seniors?

   _____

4. Fluffy's cat treat box contains 6 chicken-flavored treats, 5 beef-flavored treats, and 7 fish-flavored treats. If Fluffy's owner reaches in the box without looking, what is the probability that Fluffy will get a chicken-flavored treat?

   _____

5. Shondra used 6 ounces of chocolate chips to make two dozen cookies. At that rate, how many ounces of chocolate chips would she need to make seven dozen cookies?

   _____

6. When Rick measures the shadow of a yard stick, it is 5 inches. At the same time, the shadow of the tree he would like to chop down is 45 inches. How tall is the tree in yards?

   _____

Solve the following proportions:

7. $\frac{8}{x} = \frac{1}{2}$

8. $\frac{2}{5} = \frac{x}{10}$

9. $\frac{x}{6} = \frac{3}{9}$

10. $\frac{4}{9} = \frac{8}{x}$

11. On a scale drawing of a house floor plan, 1 inch represents 2 feet. The length of the kitchen measures 5 inches on the floor plan. How many feet does that represent?

    _____

12. If 4 inches represent 8 feet on a scale drawing, how many feet does 6 inches represent?

    _____

13. On a scale drawing, 3 centimeters represent 100 miles. If a line segment between two points measured 5 centimeters, how many miles would it represent?

    _____

14. On a map scale, 2 centimeters represent 5 kilometers. If two towns on the map are 20 kilometers apart, how long would the line segment be between the two towns on the map?

    _____

15. If 3 inches represents 10 feet on a scale drawing, how long will a line segment be that represents 15 feet?

    _____

TIME PROBLEMS

CONVERTING UNITS OF TIME

Time				Abbreviations		
1 week	=	7	days	year	=	yr
1 day	=	24	hours	week	=	wk
1 hour	=	60	minutes	hour	=	hr or h
1 minute	=	60	seconds	minutes	=	min
				seconds	=	sec

EXAMPLE: Simplify: 2 days 34 hr 75 min

Step 1: 75 minutes is more than 1 hour. There are 60 minutes in an hour, so divide 75 by 60.

$$\begin{array}{r} 1 \text{ hr} \\ 60\overline{)75} \\ -60 \\ \hline 15 \text{ min} \end{array}$$

$$\begin{array}{r} 2 \text{ days} \quad 34 \text{ hr} \quad \cancel{75} \text{ min} \\ + \quad\quad 1 \text{ hr} \quad 15 \text{ min} \\ \hline 2 \text{ days} \quad 35 \text{ hr} \quad 15 \text{ min} \end{array}$$

Step 2: 35 hours is more than 1 day. There are 24 hours in a day, so divide 35 hours by 24.

$$\begin{array}{r} 1 \text{ day} \\ 24\overline{)35} \\ -24 \\ \hline 11 \text{ hr} \end{array}$$

$$\begin{array}{r} 2 \text{ days} \quad \cancel{35} \text{ hr} \quad 15 \text{ min} \\ + 1 \text{ day} \quad 11 \text{ hr} \\ \hline 3 \text{ days} \quad 11 \text{ hr} \quad 15 \text{ min} \end{array}$$

Simplify the following.

1. 5 years 18 months

2. 4 hours 84 minutes

3. 1 minute 76 seconds

4. 3 weeks 8 days

5. 1 week 10 days

6. 3 minutes 80 seconds

7. 5 hours 75 minutes

8. 5 days 30 hours 78 min

9. 5 wk 8 days 36 hr

10. 2 hr 55 min 86 sec

11. 12 hr 86 min 87 sec

12. 7 years 13 months

ADDING UNITS OF TIME

Example: **Add 3 days 8 hours 45 minutes + 2 days 20 hours 35 minutes.**

Step 1: Arrange the numbers so that like units are in the same column.

Step 2: Add.

$$
\begin{array}{r}
3 \text{ days} \quad 8 \text{ hr} \quad 45 \text{ min} \\
+ \quad 2 \text{ days} \quad 20 \text{ hr} \quad 35 \text{ min} \\
\hline
5 \text{ days} \quad 28 \text{ hr} \quad 80 \text{ min}
\end{array}
$$

Step 3: Simplify the answer.

$$
\begin{array}{r}
5 \text{ days} \quad 28 \text{ hr} \quad 80 \text{ min} \\
\end{array}
$$

80 minutes =

$$
\begin{array}{r}
\underline{\qquad\qquad 1 \text{ hr} \quad 20 \text{ min}} \\
5 \text{ days} \quad 29 \text{ hr} \quad 20 \text{ min} \\
\end{array}
$$

29 hours =

$$
\begin{array}{r}
\underline{1 \text{ day} \quad 5 \text{ hr}} \\
6 \text{ days} \quad 5 \text{ hr} \quad 20 \text{ min}
\end{array}
$$

Add and simplify the answers.

1. 2 hr 50 min + 15 hr 15 min

2. 1 hr 20 min + 3 hr 45 min

3. 4 days 22 hr + 6 days 8 hr

4. 3 min 30 sec + 5 min 45 sec

Add and simplify the answers.

5.
$$
\begin{array}{r}
5 \text{ yr} \quad 6 \text{ months} \\
+ \ 7 \text{ yr} \ 11 \text{ months} \\
\hline
\end{array}
$$

6.
$$
\begin{array}{r}
4 \text{ days} \ 23 \text{ hr} \ 55 \text{ min} \\
+ \ 5 \text{ days} \quad 9 \text{ hr} \ 33 \text{ min} \\
\hline
\end{array}
$$

7.
$$
\begin{array}{r}
1 \text{ week} \quad 2 \text{ days} \\
+ \ 2 \text{ weeks} \ 6 \text{ days} \\
\hline
\end{array}
$$

8.
$$
\begin{array}{r}
1 \text{ yr} \quad 1 \text{ month} \\
+ \ 4 \text{ yr} \ 11 \text{ months} \\
\hline
\end{array}
$$

9.
$$
\begin{array}{r}
57 \text{ min} \ 23 \text{ sec} \\
+ \ 25 \text{ min} \ 44 \text{ sec} \\
\hline
\end{array}
$$

10.
$$
\begin{array}{r}
15 \text{ hr} \ 6 \text{ min} \\
+ \quad 7 \text{ hr} \ 5 \text{ min} \\
\hline
\end{array}
$$

SUBTRACTING UNITS OF TIME

Example: **4 hr 15 min – 2 hr 50 min**

Step 1:   Arrange the numbers so that like units are in the same column.

$$4 \text{ hr } 15 \text{ min}$$
$$- 2 \text{ hr } 50 \text{ min}$$

Step 2:   Subtract like units by borrowing when necessary.

$$\begin{array}{r} 3 \quad 75 \\ \cancel{4} \text{ hr } \cancel{15} \text{ min} \\ - 2 \text{ hr } 50 \text{ min} \\ \hline 1 \text{ hr } 25 \text{ min} \end{array}$$

You must borrow 1 hour to subtract. 1 hour = 60 minutes. You need to add 60 minutes to the 15 minutes you already have (75) and then subtract.

Subtract by borrowing when necessary.

1. 3 days 4 hr – 1 day 9 hr

2. 3 hr 16 min – 2 hr 46 min

3. 5 min 15 sec – 2 min 35 secretary

4. 4 hr 10 min – 2 hr 25 min

5. 55 min 15 sec – 22 min 43 sec

6. 3 days 12 hr – 20 hr

7.
$$2 \text{ yr } 2 \text{ mo}$$
$$- 1 \text{ yr } 8 \text{ mo}$$

8.
$$5 \text{ days } 6 \text{ hr}$$
$$- 3 \text{ days } 15 \text{ hr}$$

9.
$$5 \text{ days } 6 \text{ hr } 45 \text{ min}$$
$$- 3 \text{ days } 22 \text{ hr } 50 \text{ min}$$

10.
$$8 \text{ hr } 10 \text{ min}$$
$$- 5 \text{ hr } 45 \text{ min}$$

11.
$$11 \text{ hr } 5 \text{ min}$$
$$- 8 \text{ hr } 50 \text{ min}$$

12.
$$4 \text{ hr } 27 \text{ min}$$
$$- 2 \text{ hr } 54 \text{ min}$$

13.
$$10 \text{ min } 15 \text{ sec}$$
$$- 5 \text{ min } 25 \text{ sec}$$

14.
$$6 \text{ days } 5 \text{ hr } 15 \text{ min}$$
$$- 2 \text{ days } 10 \text{ hr } 40 \text{ min}$$

15.
$$4 \text{ week } 5 \text{ days}$$
$$- \qquad 6 \text{ days}$$

16.
$$3 \text{ hr } 20 \text{ min}$$
$$- 1 \text{ hr } 45 \text{ min}$$

17.
$$2 \text{ hr } 10 \text{ min}$$
$$- \qquad 55 \text{ min}$$

18.
$$4 \text{ hr } 5 \text{ min}$$
$$- 2 \text{ hr } 25 \text{ min}$$

CHANGING MINUTES TO HOURS

Example: **Rebecca studied 140 minutes this afternoon. How many hours is that?**

Step 1:   We need to change 140 minutes to a fraction of an hour. The top number of the fraction will be the number of minutes in the problem, 140. The bottom number is the number of minutes in an hour, 60.

$$\frac{140}{60}$$

Step 2:   Reduce the fraction. $\frac{140}{60} = 2\frac{20}{60} = 2\frac{1}{3}$

Change the following minutes to a fraction of an hour.

1. 30 min = _____ hr

2   50 min = _____ hr

3. 15 min = _____ hr

4. 20 min = _____ hr

5. 45 min = _____ hr

6. 10 min = _____ hr

7. 35 min = _____ hr

8. 55 min = _____ hr

9. 40 min = _____ hr

10. 25 min = _____ hr

11. 125 min = _____ hr

12. 130 min = _____ hr

13. 360 min = _____ hr

14. 150 min = _____ hr

15. 140 min = _____ hr

16. 75 min = _____ hr

17. 110 min = _____ hr

18. 300 min = _____ hr

19. 390 min = _____ hr

20. 85 min = _____ hr

21. 190 min = _____ hr

22. 220 min = _____ hr

23. 450 min = _____ hr

24. 205 min = _____ hr

25. 285 min = _____ hr

26. 345 min = _____ hr

27. 135 min = _____ hr

28. 280 min = _____ hr

29. 200 min = _____ hr

30. 320 min = _____ hr

31. Alisha did aerobics for 80 minutes. How many hours did she do aerobics?

32. Kitty walked for 180 minutes. How many hours did she walk?

33. Tamara read a novel for 250 minutes. How many hours did she read?

34. Eldrich read a science fiction book for 300 minutes. How many hours did he read?

35. Ross drove his truck for 430 minutes before stopping. How many hours did he drive?

CHANGING HOURS TO DIGITAL TIME

Example: Change $5\frac{2}{3}$ hours to digital time.

Step 1: Change the fraction of an hour to minutes by multiplying by 60.

$$\frac{2}{\cancel{3}} \times \frac{\cancel{60}}{1} = 40 \text{ minutes}$$

Step 2: Copy the whole number of hours followed by a colon, then copy the minutes. Example: 5:40.

Change the following hours to digital time.

1. $1\frac{1}{2}$ hr = _____

2. $2\frac{3}{4}$ hr = _____

3. $3\frac{1}{4}$ hr = _____

4. $6\frac{1}{2}$ hr = _____

5. $4\frac{1}{3}$ hr = _____

6. $6\frac{3}{4}$ hr = _____

7. $2\frac{2}{3}$ hr = _____

8. $5\frac{1}{2}$ hr = _____

9. $1\frac{1}{4}$ hr = _____

10. $4\frac{2}{3}$ hr = _____

11. $8\frac{1}{2}$ hr = _____

12. $9\frac{3}{4}$ hr = _____

13. $7\frac{1}{3}$ hr = _____

14. $2\frac{1}{10}$ hr = _____

15. $5\frac{1}{12}$ hr = _____

16. $6\frac{2}{3}$ hr = _____

17. $4\frac{1}{6}$ hr = _____

18. $5\frac{1}{3}$ hr = _____

19. $8\frac{1}{4}$ hr = _____

20. $7\frac{3}{4}$ hr = _____

21. $6\frac{4}{5}$ hr = _____

TIME WORD PROBLEMS

Example: Regina drove $2\frac{3}{4}$ hours to her sister's house in St. Cloud. She left home at 1:30 p.m. When did she arrive?

Step 1:   Write down the starting time.

Step 2:   Change hours to digital time and add. $2\frac{3}{4}$ hours = 2:45

$$\begin{array}{r} 1:30 \\ +\ 2:45 \\ \hline 3:75 \end{array}$$

Step 3:   Simplify time. 75 minutes = 1:15 so 3 + 1:15 = 4:15. She arrived at 4:15.

Answer the following word problems.

1. Kristy drove for $4\frac{1}{4}$ hours to Edina for a family reunion. She left home at 6:30 a.m. When did she arrive at the reunion?

2. Sam and his friends drove $3\frac{1}{2}$ hours to go to Mankato to pick up his cousin. They started out at 2:15 p.m. When did they arrive?

3. Brianna started work at 3:30 p.m. and stayed for $6\frac{3}{4}$ hours. When did she leave work?

4. Steve and Dad drove $2\frac{1}{6}$ hours from Deer Lake to go camping at Gooseberry Falls State Park. They left at 12:30 p.m. When did they get to their campground?

5. Kelsie and Keri left Morton at 7:30 a.m. to vacation in St. Louis. After driving $3\frac{2}{3}$ hours, they took a break. What time was it then?

6. Alicia left home at 2:30 p.m. to drive to her sister's house in Alexandria. It took her $2\frac{3}{4}$ hours. What time did she get there?

7. Kevin started work at 7:15 a.m. He worked $2\frac{3}{4}$ hours before stopping for a break. At what time did he take a break?

8. Marcia left at 1:45 p.m. to drive to Bemidji. The trip took $3\frac{1}{2}$ hours. What time did she arrive in Bemidji?

9. Joe and Ed drove to Lake Mille Lacs. They left at 10:30 a.m., and the trip took $3\frac{1}{4}$ hours. What time did they arrive?

10. Rebecca started cooking Thanksgiving dinner at 12:15 p.m. and put the turkey in the oven first. If the turkey had to cook $4\frac{1}{2}$ hours, what time was the turkey done?

TWO-STEP TIME PROBLEMS

Example: Theresa and Laura left Hibbing at 10 a.m. After $5\frac{1}{2}$ hours of driving, they arrived at their aunt's farm in North Dakota. What time was it then?

Step 1:   Compute the time from 10:00 a.m. to noon.

$$
\begin{array}{r}
12:00 \\
-\,10:00 \\
\hline
2:00 \text{ hours}
\end{array}
$$

Step 2:   Subtract these 2 hours from the total driving time to find the answer.

$$
\begin{array}{r}
5:30 \\
-\,2:00 \\
\hline
3:30 \text{ p.m.}
\end{array}
$$

They arrived at 3:30 p.m.

Answer the following word problems.

1. Trevor and Johnny drove $3\frac{3}{4}$ hours from St. Paul to go hunting. They started out at 11:30 a.m. When did they get to their campground?

2. Holly and Julie drove from their house to Mankato to visit a friend. They left at 9:00 a.m. and drove for $5\frac{1}{4}$ hours. What time was it when they arrived?

3. Mary left her home in Shakopee at 10:30 a.m. After driving $4\frac{5}{6}$ hours, she got to her brother's house. What time was it when she got there?

4. Lindsey and Allison make the $4\frac{1}{2}$ hour drive to the Mall of America every weekend during the summer. If they leave home at 11:15 a.m., what time will they get to the mall?

5. Dana and Tammy drove to their sister's house. They left at 10:30 a.m. and arrived $6\frac{3}{4}$ hours later. What time was it when they arrived?

6. Jessie left home at 4:50 a.m. to go hunting. He came back home $10\frac{1}{2}$ hours later. What time did he come home?

7. Jonathan went to work at 6:30 a.m. He got home $8\frac{1}{2}$ hours later. What time did he get home?

8. Beth went to the mall at 10:45 in the morning. She got home $6\frac{2}{3}$ hours later. What time was it when she got home?

9. Josh left home at 11:20 a.m. He returned home $7\frac{1}{2}$ hours later. What time did he get home?

10. Randy arrived at work at 9:15 a.m. He took a lunch break 4 hours later. What time did he break for lunch?

CALCULATING STARTING TIMES

Example: Doug worked $5\frac{1}{2}$ hours on Saturday. He finished work at 7:15 p.m. When did he start work?

Step 1:	Copy ending time.	6 75 ⟵—— Remember when you	
Step 2:	Change hours to time and subtract.	7̸:1̸5̸ borrow 1 hour, you	
		− 5:30 borrow 60 minutes.	
		1:45 p.m.	

Doug started work at 1:45 p.m.

Calculate the starting times in each of the problems below.

1. Kyle worked $3\frac{1}{2}$ hours cutting the lawn. He finished at 4:15 p.m. When did he start?

2. Lauren drove to Duluth in $2\frac{1}{4}$ hours. She arrived at 5:50 p.m. What time did she start driving?

3. Noah worked $5\frac{1}{4}$ hours on Thursday. He got off work at 10:30 p.m. When did he start working?

4. Ashley worked out for $2\frac{2}{3}$ hours at the gym. She finished at 4:30 p.m. When did she start?

5. The electrician took $1\frac{1}{3}$ hours to finish wiring our new air conditioner. He finished at 11:10 a.m. When did he start?

6. Jennifer works part-time in a shoe store. She finished working $3\frac{1}{2}$ hours at 5:10 p.m. When did she start?

7. Ryan spent $1\frac{1}{4}$ hours cleaning his room on Saturday morning. He finished at 11:40 a.m. When did he start?

8. Crystal and her dad spent $5\frac{1}{2}$ hours looking for a car on a Saturday. They found one to buy at 6:15 p.m. When did they start looking?

9. Zach worked $3\frac{3}{4}$ hours at the grocery store on Thursday. He got off at 8:10 p.m. When had he started working?

10. Nicole babysat for $4\frac{1}{3}$ hours last night. She was finished at 11:40 p.m. When did she start?

11. Danny worked for $3\frac{1}{6}$ hours on his car yesterday without taking a break. He finished at 5:45 p.m. When did he start?

12. Danielle and her sister drove from Brainerd to Rochester on Sunday in $5\frac{1}{4}$ hours. They arrived at 6:45 p.m. When did they leave Brainerd?

CALCULATING HOURS

Example 1: John started work at 8:15 a.m. and left at 11:45 a.m. How long did he work?

Step 1: Copy the time he left.

Step 2: Subtract the time he started.

Step 3: Change the number of hours to a mixed number.

$$\begin{array}{r} 11:45 \\ -\ 8:15 \\ \hline 3:30 \end{array} = 3\frac{1}{2} \text{ hours}$$

Example 2: Milly started work at 8:45 a.m. and left at 11:20 a.m. How long did she work?

Step 1: Copy the time she left.

Step 2: Change the work time to a mixed number.

Step 3: Change work time to a mixed number.

$$\begin{array}{r} 0\ 8 \\ 1\!\!\!1\!:\!\!\!2\!0 \\ -\ 8:45 \\ \hline 2:45 \end{array} = 2\frac{45}{60} = 2\frac{3}{4} \text{ hours}$$

Find the amount of time between starting and ending in each of the problems below. Express answers as mixed numbers.

1. Brent starts work at 3:00 p.m. and worked until 10:45 p.m. How many hours did he work?

2. Sarah left home at 6:45 a.m. and returned at 11:15 a.m. the same morning. How many hours was she gone?

3. Casey went to work at 4:15 p.m. and left at 11:55 p.m. How many hours did he work?

4. Kelly came to work at 7:30 a.m. and left to go to the dentist at 10:15 a.m. How long did she work?

5. Derek went to a movie that started at 6:30 p.m. The movie let out at 8:20 p.m. How long was the movie?

6. Clarissa went to school at 8:20 a.m. and returned home at 4:30 p.m. How long was she gone? Hint: Figure the time until 12:00 noon and then add on the 4:30.

7. Jordan went swimming at 1:10 p.m. He came home at 8:30 p.m. How long was he gone?

8. Megan went to the mall at 2:15 p.m. She got home at 4:05 p.m. How long was she gone?

9. Aaron started work at 4:15 p.m. and got out at midnight. How many hours did he work?

10. Whitney started work at 10:15 a.m. and left work at 4:30 p.m. How many hours was she at work?

11. Adam took his car in for repairs at 9:30 a.m. The car was done at 4:00 p.m. How long was his car in the shop?

12. Dawn started her homework at 5:30 p.m. and was done at 6:20 p.m. How long did it take Dawn to do her homework?

1. Jenny took 145 minutes to do her homework last night. How many hours did she work?

   _____

2. Ben mowed the lawn for $2\frac{1}{4}$ hours. He finished at 4:30 p.m. When did he start to mow the lawn?

   _____

3. Jim started driving at 6:30 a.m. After $4\frac{2}{3}$ hours on the road, he stopped for a break. What time was it then?

   _____

4. Robin went to the mall at 1:45 p.m. and stayed until closing at 9:00 p.m. How long was she there?

   _____

5. Brandee worked at the telephone company for 4 years and 11 months. Aubrey has been at the same company for 5 years and 9 months. How much longer has Aubrey worked at the company than Brandee?

   _____

6. Josh started cleaning his car at 1:20 p.m. It took him $1\frac{3}{4}$ hours to finish. At what time did he finish?

   _____

7. Abby worked at the restaurant for $6\frac{1}{2}$ hours straight on Saturday. She finished working at 8:15 p.m. When did she start work?

   _____

8. Gerald spent 220 minutes on the Internet one night. How many hours did he spend?

   _____

9. Nathan decided to go to the amusement park with his friends on Sunday. He arrived at the park at 2:30 p.m., and it took him $1\frac{2}{3}$ hours to get there. At what time did he leave his house?

   _____

10. Dannielle worked on her homework for 185 minutes. How long did she work?

   _____

DECIMALS

ROUNDING DECIMALS

EXAMPLE:

Hundreds
Tens
Ones
Tenths
Hundredths
Thousandths
Ten Thousandths

5 6 8.4 5 8 7

Consider the number 568.4587 shown with the place values labeled to the left. To round to a given place value, first find the place value in the decimal. Then look to the digit on the right. If the digit on the right is 5 or greater, INCREASE BY ONE the place value you are rounding to. All the digits to the right of the given place value are dropped if the place value is after the decimal point. If the digit on the right is LESS THAN 5, leave the place value the same. All the digits to the right of the given place value are dropped if the place value you are rounding to is after the decimal point.

Round the number 568.4587 to the nearest:

Tenth   **568.5**
Hundredth   **568.46**
Thousandth   **568.459**

Note: The decimal point is never moved when rounding.

Round to the nearest tenth.

1. 45.58 _____	4. 0.618 _____	7. 25.483 _____	10. 0.854 _____
2. 17.041 _____	5. 4.543 _____	8. 9.092 _____	11. 21.028 _____
3. 5.284 _____	6. 91.385 _____	9. 0.349 _____	12. 1.957 _____

Round to the nearest hundredth.

13. 10.049 _____	16. 32.896 _____	19. 13.006 _____	22. 31.456 _____
14. 0.753 _____	17. 6.354 _____	20. 9.965 _____	23. 4.9571 _____
15. 6.432 _____	18. 5.729 _____	21. 0.874 _____	24. 8.6274 _____

Round to the nearest thousandth.

25. 9.8457 _____	28. 8.00295 _____	31. 4.0782 _____	34. 6.5832 _____
26. 12.2854 _____	29. 7.1546 _____	32. 0.9984 _____	35. 2.7153 _____
27. 0.8542 _____	30. 5.1238 _____	33. 5.2171 _____	36. 8.39548 _____

READING AND WRITING DECIMALS

EXAMPLE: Write the number 506.402 in words.

(The digits have been lined up with their place value names in the box on the right.)

Step 1: **Write the number to the left of the decimal point:** <u>Five hundred six</u>

Step 2: **Write the word "and" where the decimal point is:** Five hundred six <u>and</u>

Step 3: **Write the numbers after the decimal point, and add the name of the column of the last digit:** Five hundred six and <u>four hundred two thousandths.</u>

Note: **Only use the word "and" where the decimal point comes. NEVER write three hundred and eight for the number 308.**

Write the following numbers in words.

1.   951.32 _____

2.   542.236 _____

3.   10.058 _____

4.   901.47 _____

5.   5.0247 _____

6.   54.0587 _____

7.   23.503 _____

8.   65.0245 _____

9.   7.653 _____

10.   328.473 _____

11.   80.057 _____

12.   0.548 _____

13.   570.007 _____

14.   95.0345 _____

CHECK WRITING

Checks are used to purchase merchandise or services in stores, offices, or through the mail. They are widely used because checks are a safe and efficient way to pay for goods and services without risking the theft of money when you use cash.

When filling out a check, make sure you write the check correctly. Below are the six parts that must be filled out when writing a check. Each number corresponds to one part of the check.

1. **The Date -** In this blank, you will write today's date. First, write the month and day, and then write out the year.

2. **The Payee -** After the phrase **"Pay To The Order Of,"** write the name of the person or institution that will receive the money; for example, Dan's Machine Shop.

3. **The Amount -** In this blank or box, write the amount of money in numerals; for example, $68.35.

4. **The Written Amount -** On this line, write the amount of money by spelling out the dollar amount in words. For the cents, write the amount divided by 100; for example, write Sixty-eight and 35/100.

5. **Memo -** Next to the word **FOR** on the check, write a note that describes the type of purchase you made; for example, car repair.

6. **Signature -** Sign your name on the next line on the right. Make your signature unique so that it will be difficult for someone to forge. Use both your first and last name when writing your signature.

First, practice writing out the dollar amount.

EXAMPLE 1: On a check written for $12.60, write out: Twelve and $\frac{60}{100}$

In this example, the number of dollars is spelled out (twelve). The word "and" is where the decimal point is. The cents are expressed as a fraction with 60 as the numerator (top number) and 100 as the denominator (bottom number). If there are no cents, write the word "no" over 100.

OTHER EXAMPLES:

DOLLAR AMOUNT	WRITTEN IN WORDS WITH CENTS AS A FRACTION
$ 53.41	Fifty-three and $\frac{41}{100}$
$120.70	One hundred twenty and $\frac{70}{100}$
$504.00	Five hundred four and $\frac{no}{100}$

Write the following dollar amounts in words with cents expressed as a fraction.

1. $89.31 _____

2. $105.30 _____

3. $54.36 _____

4. $879.00 _____

5. $45.77 _____

6. $1,980.00 _____

7. $4,789.14 _____

8. $29.87 _____

9. $365.23 _____

10. $1,431.00 _____

11. $563.41 _____

12. $639.50 _____

EXAMPLE 2: John McCoy went to Dan's Machine Shop for some car repairs on May 26, 2004. The bill was for $68.35. John wrote a check for the charges. The check below is made out correctly.

A check showing:

2001

May 26   2004

PAY TO THE ORDER OF   Dan's Machine Shop $ 68.35

Sixty-eight and 35/100   DOLLARS

FOR   car repair   John McCoy

Fill out the checks below using the information provided.

1. Jackie Goodrum is purchasing a lawn mower from Exterior Central for her landscaping business. Today is June 2, 2004. The lawn mower costs $376.45 with tax included.

2. Betsy Wheeler is buying groceries at Jay's Food Mart. Today is March 5, 2004. Her total grocery bill is $54.64.

ADDING DECIMALS

EXAMPLE: Find 0.9 + 2.5 + 63.17

Step 1: When you add decimals, first arrange
the numbers in columns with
the decimal points under each other.

$$\begin{array}{r} 0.9 \\ 2.5 \\ + \ 63.17 \end{array}$$

Step 2: Add 0's here to keep your columns straight.

Step 3: Start at the right and add each column.
Remember to carry when necessary.
Bring down the decimal point.

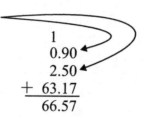

$$\begin{array}{r} 1 \\ 0.90 \\ 2.50 \\ + \ 63.17 \\ \hline 66.57 \end{array}$$

Add. Be sure to write the decimal point in your answer.

1. 5.3 + 6.02 + 0.73

2. 0.235 + 6.2 + 3.27

3. 7.542 + 10.5 + 4.57

4. $5.87 + $7.52

5. $4.68 + $9.47

6. 5.08 + 11.2 + 6.075

7. 5.14 + 2.3 + 5.097

8. 4.9 + 15.71 + 0.254

9. $3.75 + $18.90

10. $64.95 + $4.63

11. 1.25 + 4.1 + 10.007

12. 15.4 + 5.074 + 3.15

13. 45.23 + 9.5 + 0.693

14. $8.63 + $ 12.50

15. $6.87 + $27.23

16. 0.23 + 5.9 + 12

17. 8.5784 + 10.03

18. 85.7 + 205.952

19. $98.45 + $8.89

20. $7.77 + $11.19

SUBTRACTING DECIMALS

EXAMPLE: Find $14.9 - 0.007$

Step 1: When you subtract decimals, arrange
the numbers in columns with the decimal
points under each other.

$$\begin{array}{r} 14.9 \\ -\ 0.007 \\ \hline \end{array}$$

Step 2: You must fill in the empty places with 0's
so that both numbers have the same number
of digits after the decimal point.

$$\begin{array}{r} 14.900 \\ -\ 0.007 \\ \hline \end{array}$$

Step 3: Start at the right, and subtract each column.
Remember to borrow when necessary.

$$\begin{array}{r} {}^{89}14.900 \\ -\ 0.007 \\ \hline 14.893 \end{array}$$

Subtract. Be sure to write the decimal point in your answer.

1. $5.25 - 4.7$

2. $23.657 - 9.83$

3. $\$56.54 - \$17.92$

4. $\$294.78 - \$80.99$

5. $\$70.00 - \$68.99$

6. $58.6 - 9.153$

7. $405.97 - 7.325$

8. $\$40.09 - \$9.99$

9. $\$115.45 - \$4.79$

10. $\$45.18 - \$23.65$

11. $12.96 - 7.32$

12. $19.2 - 8.63$

13. $8.123 - 5.096$

14. $\$14.32 - \$0.58$

15. $\$30.00 - \$22.95$

16. $15.789 - 6.32$

17. $478.63 - 99.2$

18. $\$15.45 - \$8.58$

19. $102.5 - 1.079$

20. $7.054 - 3.009$

DETERMINING CHANGE

EXAMPLE: Jamie bought 2 T-shirts for $13.95 each and paid $1.68 sales tax. How much change should Jamie get from a $50.00 bill?

Step 1: Find the total cost of items and tax.

$$\begin{array}{r} \$13.95 \\ 13.95 \\ +\ 1.68 \\ \hline \$29.58 \end{array}$$

Step 2: Subtract the total cost from the amount of money given.

$$\begin{array}{r} \$50.00 \\ -\ 29.58 \\ \hline \end{array}$$

Change ⟶ $20.42

Find the correct change for each of the following problems.

1. Kenya bought a leather belt for $22.89 and a pair of earrings for $4.69. She paid $1.38 sales tax. What was her change from $30.00?

2. Mark spent a total of $78.42 on party supplies. What was his change from a $100.00 bill?

3. The Daniels spent $42.98 at a steak restaurant. How much change did they receive from $50.00?

4. Myra bought a sweater for $49.95 and a dress for $85.89. She paid $9.51 in sales tax. What was her change from $150.00?

5. Roland bought a calculator for $22.78 and an extra battery for $5.69. He paid $1.56 sales tax. What was his change from $40.00?

6. For lunch, Daul purchased 2 hotdogs for $1.09 each, a bag of chips for $0.89, and a large drink for $1.39. He paid $0.18 sales tax. What was his change from $10.00?

7. Geri bought a dining room set for $2,265.99. She paid $135.96 sales tax. What was her change from $2500.00?

8. Juan purchased a bag of dog food for $5.89, a leash for $11.88, and a dog collar for $4.75. The sales tax on the purchase was $1.13. How much change did he get back from 25.00?

9. Maxine bought a blouse for $15.46 and a shirt for $23.58. She paid $3.12 sales tax. What was her change from $50.00?

10. Jackie paid for four houseplants that cost $4.95 each. She paid $1.19 sales tax. How much change did she receive from 30.00?

11. Bo spent a total of $13.59 on school supplies. How much change did he receive from $14.00?

12. Fran bought 4 packs of candy on sale for 2 for $0.99 and 2 sodas for $0.65 each. She paid $0.13 sales tax. What was her change from a $5.00 bill?

MULTIPLICATION OF DECIMALS

EXAMPLE: 56.2×0.17

Step 1: Set up the problem as if you
were multiplying whole numbers.

$$\begin{array}{r} 56.2 \\ \times\ 0.17 \end{array}$$

Step 2: Multiply as if you were multiplying
whole numbers.

$$\begin{array}{r} {}^{4\ 1} \\ 56.2 \\ \times\ 0.17 \\ \hline {}^{1}3934 \\ 562 \\ \hline 9.554 \end{array}$$

56.2 ← 1 number after the decimal point
× 0.17 ← + 2 numbers after the decimal point
3 numbers after the decimal point

Step 3: Count how many numbers are
after the decimal points in the problem.
In this problem 2, 1, and 7 come after
the decimal points, so the answer must
also have three numbers after the decimal point.

Multiply.

1. 15.2×3.5

2. 9.54×5.3

3. 5.72×6.3

4. 4.8×3.2

5. 45.8×2.2

6. 4.5×7.1

7. 0.052×0.33

8. 4.12×6.8

9. 23.65×9.2

10. 1.54×0.43

11. 0.47×6.1

12. 1.3×1.57

13. 16.4×0.5

14. 0.87×3.21

15. 5.94×0.65

16. 7.8×0.23

MORE MULTIPLYING DECIMALS

EXAMPLE: Find 0.007 × 0.125

Step 1:   Multiply as you would whole numbers.

$$0.007 \longleftarrow \text{3 numbers after the decimal point}$$
$$\underline{\times\ 0.125} \longleftarrow +\ \underline{\text{3 numbers after the decimal point}}$$
$$0.000875 \longleftarrow \text{6 numbers after the decimal point}$$

Step 2:   Count how many numbers are behind decimal points in the problem. In this case, 6 numbers come after decimal points in the problem, so there must be 6 numbers after the decimal point in the answer. In this problem, 0's needed to be written in the answer **in front of** the 8 so there will be 6 numbers after the decimal point.

Multiply.

1.   0.123 × .45

2.   0.004 × 10.31

3.   1.54 × 1.1

4.   10.05 × 0. 45

5.   9.45 × 0.8

6. $6.49 × 0.06

7.   5.003 × 0.009

8. $9.99 × 0.06

9.   6.09 × 5.3

10. $22.00 × 0.075

11.   5.914 × 0.02

12.   4.96 × 0.23

13.   6.98 × 0.02

14.   3.12 × 0.08

15.   7.158 × 0.09

16.   0.0158 × 0.32

GROSS PAY

Gross pay is the amount you earn before taxes, insurance, and other deductions are taken out.

EXAMPLE: Codie earns $8.50 per hour. Last week he worked 38 hours. What was his gross pay?

Solution: Multiply the pay per hour by the number of hours worked.

$$\begin{array}{r} \$8.50 \\ \times\ 38 \\ \hline 6800 \\ 2550 \\ \hline \$323.00 \end{array}$$

Find the gross pay (total earnings before deductions) in each of the following problems.

1. Ron earns $8.25 per hour and works 35 hours each week. How much does he earn per week?

2. Casie earns $13.00 per hour and worked 40 hours last week. How much did she earn last week?

3. Maria earns $6.75 an hour at her part-time job. Last week she worked 15 hours. How much did she earn?

4. Roby worked 22.5 hours last week, and he earns $8.60 per hour. How much was his gross pay?

5. Tikki worked 11 hours at her job that pays $9.15 per hour. How much did she earn?

6. Murray worked 35 hours last week and makes $6.45 per hour. What was his gross pay?

7. Paula's job pays $12.00 per hour. Last week she worked 11.25 hours. How much did she earn last week?

8. Taylor earns $6.50 per hour working in a fast food restaurant. If he works 23 hours per week, what is his gross pay per week?

9. Mark earns $9.50 per hour painting houses. Last week he worked 36 hours. How much did he earn?

10. Kirby works in a greenhouse for $8.75 per hour. He works 40 hours per week. What is his gross pay each week?

11. Julie earns $25.00 per hour teaching tennis part time. If she works 8.5 hours in a week, how much will she earn?

12. Yvonne works in a boutique for 25 hours per week. The boutique pays her $7.15 per hour. How much does she earn each week?

DIVISION OF DECIMALS BY WHOLE NUMBERS

EXAMPLE: $52.26 \div 6$

Step 1: Copy the problem as you would for whole numbers. Copy the decimal point directly above in the place for the answer.

$$6 \overline{)52.26}$$

Step 2: Divide the same way as you would with whole numbers.

$$
\begin{array}{r}
8.71 \\
6\overline{)52.26} \\
-48 \\
\hline
4\,2 \\
-4\,2 \\
\hline
6 \\
-6 \\
\hline
0
\end{array}
$$

Divide. Remember to copy the decimal point directly above the place for the answer.

1. $42.75 \div 3$ 5. $12.50 \div 2$ 9. $72.36 \div 4$ 13. $102.5 \div 5$

2. $74.16 \div 6$ 6. $224.64 \div 52$ 10. $379.5 \div 15$ 14. $113.4 \div 9$

3. $81.50 \div 25$ 7. $183.04 \div 52$ 11. $152.25 \div 21$ 15. $585.14 \div 34$

4. $82.46 \div 14$ 8. $281.52 \div 23$ 12. $40.375 \div 19$ 16. $93.6 \div 24$

CHANGING FRACTIONS TO DECIMALS

EXAMPLE: Change $\frac{1}{8}$ to a decimal.

Step 1: To change a fraction to a decimal, simply divide the top number by the bottom number.

$$8\overline{)1}$$

Step 2: Add a decimal point and a 0 after the 1 and divide.

$$\begin{array}{r} 0.\,1 \\ 8\overline{)1.\,0} \\ -8 \\ \hline 2 \end{array}$$

Step 3: Continue adding 0's and dividing until there is no remainder.

$$\begin{array}{r} 0.\,1\,2\,5 \\ 8\overline{)1.\,0\,0\,0} \\ -8 \\ \hline 2\,0 \\ -1\,6 \\ \hline 4\,0 \\ -4\,0 \\ \hline 0 \end{array}$$

In some problems, the number after the decimal point begins to repeat. Take, for example, the fraction $\frac{4}{11}$. $4 \div 11 = 0.363636$, and the 36 keeps repeating forever. To show that the 36 repeats, simply write a bar above the numbers that repeat, $0.\overline{36}$.

Change the following fractions to decimals.

1. $\frac{4}{5}$ 5. $\frac{1}{10}$ 9. $\frac{3}{5}$ 13. $\frac{7}{9}$ 17. $\frac{3}{16}$

2. $\frac{2}{3}$ 6. $\frac{5}{8}$ 10. $\frac{7}{10}$ 14. $\frac{9}{10}$ 18. $\frac{3}{4}$

3. $\frac{1}{2}$ 7. $\frac{5}{6}$ 11. $\frac{4}{11}$ 15. $\frac{1}{4}$ 19. $\frac{8}{9}$

4. $\frac{5}{9}$ 8. $\frac{1}{6}$ 12. $\frac{1}{9}$ 16. $\frac{3}{8}$ 20. $\frac{5}{12}$

CHANGING MIXED NUMBERS TO DECIMALS

If there is a whole number with a fraction, write the whole number to the left of the decimal point. Then change the fraction to a decimal.

EXAMPLES: $\quad 4\frac{1}{10} = 4.1 \qquad 16\frac{2}{3} = 16.\overline{6} \qquad 12\frac{7}{8} = 12.875$

Change the following mixed numbers to decimals.

1. $5\frac{2}{3}$

2. $8\frac{5}{11}$

3. $15\frac{3}{5}$

4. $13\frac{2}{3}$

5. $30\frac{1}{3}$

6. $3\frac{1}{2}$

7. $1\frac{7}{8}$

8. $4\frac{9}{100}$

9. $6\frac{4}{5}$

10. $13\frac{1}{2}$

11. $12\frac{4}{5}$

12. $11\frac{5}{8}$

13. $7\frac{1}{4}$

14. $12\frac{1}{3}$

15. $1\frac{5}{8}$

16. $2\frac{3}{4}$

17. $10\frac{1}{10}$

18. $20\frac{2}{5}$

19. $4\frac{9}{10}$

20. $5\frac{4}{11}$

96

CHANGING DECIMALS TO FRACTIONS

EXAMPLE: 0.25

Step 1: Copy the decimal without the point. This will be the top number of the fraction.

$$\frac{25}{\Box}$$

Step 2: The bottom number is a 1 with as many 0's after it as there are digits in the top number.

$\frac{25}{100}$ ← Two digits ← Two 0's

$$\frac{25}{100} = \frac{1}{4}$$

Step 3: You then need to reduce the fraction.

EXAMPLES: $.2 = \frac{2}{10} = \frac{1}{5}$ $.65 = \frac{65}{100} = \frac{13}{20}$ $.125 = \frac{125}{1000} = \frac{1}{8}$

Change the following decimals to fractions.

1. .55	5. .75	9. .71	13. .35
2. .6	6. .82	10. .42	14. .96
3. .12	7. .3	11. .56	15. .125
4. .9	8. .42	12. .24	16. .375

CHANGING DECIMALS WITH WHOLE NUMBERS TO MIXED NUMBERS

EXAMPLE: Change 14.28 to a mixed number.

Step 1: Copy the portion of the number that is whole.   14

Step 2: Change .28 to a fraction. $14\frac{28}{100}$

Step 3: Reduce the fraction. $14\frac{28}{100} = 14\frac{7}{25}$

Change the following decimals to mixed numbers.

1. 7.125	5. 16.95	9. 6.7	13. 13.9
2. 99.5	6. 3.625	10. 45.425	14. 32.65
3. 2.13	7. 4.42	11. 15.8	15. 17.25
4. 5.1	8. 15.84	12. 8.16	16. 9.82

DIVISION OF DECIMALS BY DECIMALS

EXAMPLE: $374.5 \div 0.07$

Step 1: Copy the problem as you would for whole numbers.

Divisor
$0.07 \overline{)374.5}$ ← **Dividend**

Step 2: You cannot divide by a decimal number. You must move the decimal point in the divisor 2 places to the right to make it a whole number. The decimal point in the dividend must also move to the right the same number of places. Notice you must add a 0 to the dividend.

$0.07 \overline{)374.50.}$

Step 3: The problem now becomes $37450 \div 7$. Copy the decimal point from the dividend straight above in the place for the answer.

```
         5350.
007.)37450.
    -35
     24
    -21
     35
    -35
     00
```

Divide. Remember to move the decimal points.

1. $0.676 \div 0.013$

5. $18.46 \div 1.3$

9. $154.08 \div 1.8$

13. $4.8 \div 0.08$

2. $70.32 \div 0.08$

6. $14.6 \div 0.002$

10. $0.4374 \div 0.003$

14. $1.2 \div 0.024$

3. $\$54.60 \div 0.84$

7. $\$125.25 \div 0.75$

11. $292.9 \div 0.29$

15. $15.725 \div 3.7$

4. $\$10.35 \div 0.45$

8. $\$33.00 \div 1.65$

12. $6.375 \div 0.3$

16. $\$167.50 \div 0.25$

ESTIMATING DIVISION OF DECIMALS

EXAMPLE: The following division of decimals problem has the decimal point missing from the answer. Estimate the answer to determine where the decimal point should go.

$$2489 \div 5.8 = 42913793$$

Step 1: Round off the numbers in the problem to numbers that are divisible without a remainder.

5.8 rounds to 6
2489 rounds to 2400

$$2400 \div 6 = 400$$

Step 2: The answer should be close to 400, so put the decimal point after the third whole number.

Solution: $2489 \div 5.8 = 429.13793$

For each of the following problems, round off the numbers to determine where the decimal point belongs in the answer.

1. $15.63 \div 4.2 = 3\ 7\ 2\ 1\ 4$

2. $476.3 \div 5.81 = 8\ 1\ 9\ 7\ 9$

3. $7561.5 \div 10.6 = 7\ 1\ 3\ 3\ 4\ 9$

4. $6259 \div 8.1 = 7\ 7\ 2\ 7\ 1\ 6$

5. $11.78 \div .94 = 1\ 2\ 5\ 3\ 1\ 9$

6. $45.69 \div 4.67 = 9\ 7\ 8\ 3\ 7$

7. $768 \div 22.35 = 3\ 4\ 3\ 6\ 2\ 4$

8. $5.16 \div 1.78 = 2\ 8\ 9\ 8\ 8\ 7$

9. $87.32 \div 56.7 = 1\ 5\ 4\ 0\ 0\ 3\ 5$

10. $144.92 \div 12.4 = 1\ 1\ 6\ 8\ 7\ 0\ 9$

11. $456.98 \div 21.5 = 2\ 1\ 2\ 5\ 4\ 8\ 8$

12. $19 \div 8.6 = 2\ 2\ 0\ 9\ 3$

13. $79.19 \div 7.8 = 1\ 0\ 1\ 5\ 2\ 5\ 6$

14. $856.3 \div 8.2 = 1\ 0\ 4\ 4\ 2\ 6\ 8$

15. $11.235 \div .48 = 2\ 3\ 4\ 0\ 6$

16. $9.63 \div 4.1 = 2\ 3\ 4\ 8\ 7\ 8$

17. $96.68 \div 32.56 = 2\ 9\ 6\ 9\ 2\ 8\ 7$

18. $162.3 \div 87.5 = 1\ 8\ 5\ 4\ 8\ 5\ 7$

19. $45.98 \div 2.9 = 1\ 5\ 8\ 5\ 5$

20. $32.65 \div 1.689 = 1\ 9\ 3\ 3\ 0\ 9\ 7$

21. $26.5 \div 5.1 = 5\ 1\ 9\ 6$

22. $6.59 \div 2.147 = 3\ 0\ 6\ 9\ 3\ 9\ 9$

23. $75.26 \div 8.36 = 9\ 0\ 0\ 2\ 3\ 9$

24. $158.4 \div 3.09 = 5\ 1\ 2\ 6\ 2$

BEST BUY

When products come in different sizes, you need to figure out the cost per unit to see which is the best buy. Often the box marked "economy size" is not really the best buy.

EXAMPLE:

Smithfield's Instant Coffee comes in three sizes. Which one has the lowest cost per unit? The coffee comes in 8, 12, and 16 ounce sizes. To figure the lowest cost per unit, you need to see how much each unit, in this case ounce, costs in each size. If 8 ounces of coffee costs $3.60, then 1 ounce costs $3.60 ÷ 8 or $.45. $.45 is the unit cost, the cost of 1 ounce. We need to figure the unit cost for each size:

SMITHFIELD'S
Instant Coffee

8 ounces $3.60
12 ounces $5.52
16 ounces $7.44

$3.60 ÷ 8 = $.45
$5.52 ÷ 12 = $.46
$7.44 ÷ 16 = $.465

The 8 ounce size is the best one to buy because it has the lowest cost per unit.

Figure the unit cost of each item in each question below to find the best buy.
Underline the answer.

1. Which costs the **most** per ounce: 60 ounces of peanut butter for $5.40, 28 ounces for $2.24, or 16 ounces for $1.76?

2. Which is the **least** per pound: 5 pounds of chicken for $9.45, 3 pounds for $5.97, or 1 pound for $2.05?

3. Which costs the most per disk: a 10-pack of $3\frac{1}{2}$ inch floppy disks for $5.99, a 25-pack for $12.50, or a 50-pack for $18.75 ?

4. Which is the best buy: 6 ballpoint pens for $4.80 or 8 for $6.48?

5. Which costs the least per ounce: a 20-ounce soda for $0.60, 68 ounces for $2.38, or 100 ounces for $3.32?

6. Which costs more: oranges selling at 3 for $1.00 or oranges selling 4 for $1.36?

7. Which is the best buy: 1 roll of paper towels for $2.13, 3 rolls for $5.88, or 15 rolls for $29.55?

8. Which costs the most per tablet: 50 individually wrapped pain reliever tablets for $9.50, 100 tablets in a bottle for $6.32, or 500 tablets in a bottle for $13.42?

9. Which costs the least per can: a 24 pack of cola for $5.52, a 12 pack of cola for $2.64, or a 6 pack of cola for $1.35?

10. Which costs less per bag: 18 tea bags for $2.70 or 64 tea bags for $9.28?

11. Which is the best buy: a 3-pack of correction fluid for $2.97 or a 12-pack for $11.76?

12. Which is the least per roll: 1 roll of masking tape for $2.45, a 3-roll pack for $7.38, or a 12-roll pack for $29.16?

100

BEST BUY SAVINGS

EXAMPLE:	Item:	Store A	Store B
	gallon of paint	$15.99	$17.89

To find the savings, subtract the lower price from the higher price.

$17.89
− $15.99
savings ⟶ $ 1.90

Find the amount of savings by choosing the lower price in each of the following. Assume items are identical in each store.

Item	Store A	Store B	Savings
1. 6-pack of soda	$2.49	$3.12	_____
2. alarm clock	$15.95	$12.79	_____
3. space heater	$23.85	$25.97	_____
4. sports car	$25,899	$29,489	_____
5. video game	$19.99	$18.77	_____
6. garden shovel	$7.89	$15.39	_____
7. mountain bike	$596	$603	_____
8. goldfish	$0.89	$0.56	_____
9. large screen TV	$1579	$1245	_____
10. roller blades	$116.95	$121.79	_____
11. cotton shorts	$16.99	$12.79	_____
12. oriental rug	$152.86	$169.45	_____
13. wrist watch	$19.59	$16.88	_____
14. gold hoop earrings	$48.75	$56.79	_____

ORDERING DECIMALS

EXAMPLE: Order the following decimals from greatest to least.

.3, .029, .208, .34

Step 1: Arrange numbers with decimal points directly under each other.

.3
.029
.208
.34

Step 2: Fill in with 0's so they all have the same number of places after the decimal point.

.300

Read the numbers as if the decimal points were not there.

.029 ⟵ **Least**
.208
.340 ⟵ **Greatest**

Answer: .34, .3, .208, .029

Order each set of decimals below from greatest to least.

1. .075, .705, .7, .75

2. .5, .56, .65, .06

3. .9, .09, .099, .95

4. .6, .59, .06, .66

5. .3, .303, .03, .33

6. .02, .25, .205, .5

7. .004, .44, .045, .4

8. .59, .905, .509, .099

9. .1, .01, .11, .111

10. .87, .078, .78, .8

11. .41, .45, .409, .49

12. .754, .7, .74, .75

13. .63, .069, .07, .06

14. .23, .275, .208, .027

Order each set of decimals below from least to greatest.

15. .055, .5, .59, .05

16. .7, .732, .74, .72

17. .04, .48, .048, .408

18. .9, .905, .95, .09

19. .19, .09, .9, .1

20. .21, .02, .021, .2

21. .038, .3, .04, .38

22. .695, .59, .065, .69

23. .08, .88, .808, .008

24. .015, .05, .105, .15

25. .4, .407, .47, .047

26. .632, .63, .603, .62

27. .02, .022, .222, .20

28. .541, .54, .504, .5

DECIMAL WORD PROBLEMS

1. Micah can have his oil changed in his car for $19.99, or he can buy the oil and filter and change it himself for $8.79. How much would he save by changing the oil himself?

        _____

2. Megan bought 5 boxes of cookies for $3.75 each. How much did she spend?

        _____

3. Will subscribes to a monthly auto magazine. His one year subscription cost $29.97. If he pays for the subscription in 3 equal installments, how much is each payment?

        _____

4. Pat purchases 2.5 pounds of hamburger at $0.98 per pound. What is the total cost of the hamburger?

        _____

5. The White family took $650 cash with them on vacation. At the end of their vacation, they had $4.67 left. How much cash did they spend on vacation?

        _____

6. Acer Middle School spent $1,443.20 on 55 math books. How much did each book cost?

        _____

7. The Junior Beta Club needs to raise $1,513.75 to go to a national convention. If they decide to sell candy bars at $1.25 each, how many will they to need to sell to meet their goal?

        _____

8. Fleta owns a candy store. On Monday, she sold 6.5 pounds of chocolate, 8.34 pounds of jelly beans, 4.9 pounds of sour snaps, and 5.64 pounds of yogurt-covered raisins. How many pounds of candy did she sell total?

        _____

9. Randal purchased a rare coin collection for $1,803.95. He sold it at auction for $2,700. How much money did he make on the coins?

        _____

10. A leather jacket that normally sells for $259.99 is on sale now for $197.88. How much can you save if you buy it now?

        _____

11. At the movies, Gigi buys 0.6 pounds of candy priced at $2.10 per pound. How much did she spend on candy?

        _____

12. George has $6.00 to buy candy. If each candy bar costs $.60, how many bars can he buy?

        _____

13. While training for a marathon, Todd runs 12.5 miles on Monday, 15.6 miles on Tuesday, 19.25 miles on Wednesday, 20.45 miles on Thursday, and 13 miles on Friday. How many miles did he run Monday through Friday?

_____

14. Thomas worked 35 hours and earned $288.75 in gross pay. How much does he make per hour?

_____

15. Kathy earns a salary plus commission. Her base salary is $145.25 per week. If last week her gross pay was $384.12, how much of her pay was from commission?

_____

16. A copy shop leases a copy machine for $256.25 per month plus $0.008 per copy made on the machine. If the copy shop makes 12,480 copies in a month, what will be the lease payment?

_____

17. Mike purchased a CD for $12.95, batteries for $7.54, and earphones for $24.68. He paid $2.94 in sales tax. How much did he spend?

_____

18. If gasoline costs $1.23 per gallon, how much would it cost to buy 12.5 gallons?

_____

19. Kevin is saving money to buy a TV that sells for $350.75. He earns money by cutting his neighbors' lawns for $15.25 per lawn. How many lawns must he cut before he will have enough money to buy the TV?

_____

20. T-shirts were donated at no cost to the Barker Valley High School Band. The band then sold 278 of the T-shirts at $8.75 each to raise money. How much money did they raise?

_____

21. Neil drives 23.48 miles to work each day and the same distance back home. How many total miles does he travel to work and back Monday through Friday?

_____

22. When Jude travels on company business using his own car, his company reimburses him $0.26 per mile. How much will Jude get reimbursed if he travels 352.5 miles?

_____

104

CHECKBOOK REGISTERS

Example: What is the balance after purchasing groceries at Kelly's Grocery?

DATE		CHECK NUMBER	PAYEE OF CHECK/ TRANSACTION DESCRIPTION	(+) AMOUNT OF DEPOSIT		(-) AMOUNT OF CHECK		BALANCE 147.82	
4	9	1148	Kelly's Grocery			65	36	?	?

Step 1: Copy the balance. $147.82 ← This is the amount in your bank.
Step 2: Subtract the amount of the check - $ 65.36
  $ 82.46 ← This is the new balance.

Find the new balance after each transaction below.

1.

DATE		CHECK NUMBER	PAYEE OF CHECK/ TRANSACTION DESCRIPTION	(+) AMOUNT OF DEPOSIT		(-) AMOUNT OF CHECK		BALANCE 498.92	
5	21	167	Southern Electric			89	70		

2.

DATE		CHECK NUMBER	PAYEE OF CHECK/ TRANSACTION DESCRIPTION	(+) AMOUNT OF DEPOSIT		(-) AMOUNT OF CHECK		BALANCE 179.60	
10	2	419	T Mart			114	20		

3.

DATE		CHECK NUMBER	PAYEE OF CHECK/ TRANSACTION DESCRIPTION	(+) AMOUNT OF DEPOSIT		(-) AMOUNT OF CHECK		BALANCE 512.79	
8	4	598	Smith Electronics			79	40		

4.

DATE		CHECK NUMBER	PAYEE OF CHECK/ TRANSACTION DESCRIPTION	(+) AMOUNT OF DEPOSIT		(-) AMOUNT OF CHECK		BALANCE 488.76	
6	21	1172	Sandy's Hair Salon			32	25		

5.

DATE		CHECK NUMBER	PAYEE OF CHECK/ TRANSACTION DESCRIPTION	(+) AMOUNT OF DEPOSIT		(-) AMOUNT OF CHECK		BALANCE 694.32	
5	12	6490	Myers Tires			114	70		

6.

DATE		CHECK NUMBER	PAYEE OF CHECK/ TRANSACTION DESCRIPTION	(+) AMOUNT OF DEPOSIT		(-) AMOUNT OF CHECK		BALANCE 719.75	
7	8	691	Jack's Hardware			159	88		

7.

DATE		CHECK NUMBER	PAYEE OF CHECK/ TRANSACTION DESCRIPTION	(+) AMOUNT OF DEPOSIT		(-) AMOUNT OF CHECK		BALANCE 498.34	
12	19	490	Trina's Trinkets			15	81		

MORE CHECKBOOK REGISTERS

Example: What is the balance after the following purchase and deposit?

DATE		CHECK NUMBER	PAYEE OF CHECK/ TRANSACTION DESCRIPTION	(+) AMOUNT OF DEPOSIT		(-) AMOUNT OF CHECK		BALANCE 689.40	
1	14	477	Southern Gas Company			122	70	?	?
1	15		Deposit	419	25			?	?

Step 1: Subtract the amount of the check →

$$\begin{array}{r} \$689.40 \\ - \ \$122.70 \\ \hline \$566.70 \end{array}$$

Step 2: Add deposit amount to new balance →

$$\begin{array}{r} + \ \$419.25 \\ \hline \$985.95 \end{array}$$

Find the new balance after each transaction below.

1.

DATE		CHECK NUMBER	PAYEE OF CHECK/ TRANSACTION DESCRIPTION	(+) AMOUNT OF DEPOSIT		(-) AMOUNT OF CHECK		BALANCE 249.15	
8	19	398	Intelligent Software			34	95		
8	21		Deposit	420	72				

2.

DATE		CHECK NUMBER	PAYEE OF CHECK/ TRANSACTION DESCRIPTION	(+) AMOUNT OF DEPOSIT		(-) AMOUNT OF CHECK		BALANCE 80.61	
2	29		Deposit	460	96				
3	2	472	Mindy's Jewelry			112	19		

3.

DATE		CHECK NUMBER	PAYEE OF CHECK/ TRANSACTION DESCRIPTION	(+) AMOUNT OF DEPOSIT		(-) AMOUNT OF CHECK		BALANCE 122.02	
10	27	814	Eagle's Grocery			64	13		
10	27		Mindy's Jewelry	811	12				

4.

DATE		CHECK NUMBER	PAYEE OF CHECK/ TRANSACTION DESCRIPTION	(+) AMOUNT OF DEPOSIT		(-) AMOUNT OF CHECK		BALANCE 614.07	
5	5	1124	Pet Paradise			42	20		
5	7		Deposit	112	14				

5.

DATE		CHECK NUMBER	PAYEE OF CHECK/ TRANSACTION DESCRIPTION	(+) AMOUNT OF DEPOSIT		(-) AMOUNT OF CHECK		BALANCE 614.07	
9	9		Deposit	272	90				
9	10	215	Town Cleaners			16	50		

Answer the following questions. Be sure to subtract checks and bank charges, and add deposits to the original balance to find the current balance.

1. Patti had a balance of $273.33 on June 3. On June 10, she wrote checks for $26.89 and $106.47. On June 15, she deposited $349.20. On June 28, she paid a $2.25 service charge. What was her balance at the end of the month?

2. Lucas started October with a balance of $726.82. He made a $248.90 truck payment, wrote checks of $7.89 and $426.10, made a $240.00 deposit, and paid $4.50 in bank charges. What was his balance at the end of the month?

3. Angela had a balance of $512.14 on May 1. She deposited $243.84 on May 9, and later that month wrote checks for $612.18 and $14.91. Her bank charged a $5.00 service fee. What was her closing balance for the month of May?

4. Ty had $144.26 in his checking account. He was charged a $10.00 service fee, deposited $209.41, and wrote two checks for $91.14 and $56.87. What was his balance after these transactions?

5. Charlene had a balance of $192.14 on July 1. She deposited $412.19 on July 5. She wrote checks for $56.90 and $17.34. Her bank service charge was $5.00. What was her balance after these transactions?

6. Bobby started January with a balance of $1.29. He deposited $54.98. He wrote checks of $17.50 and $15.40. His service charge from the bank was $7.00. What was his new balance after these transactions?

7. Corey started the month of May with a balance of $17.95. He made a deposit of $149.80. His bank service charge was $2.50, and he wrote checks for $30.95 and $55.90. What was his balance after these transactions?

8. Sheila opened a checking account with $187.00. She wrote a check for $83.90, deposited $280.95, and wrote a check for $29.15. What was her balance after these transactions?

9. Shakah had $319.19 in his checking account on April 1. He wrote a check for $214.62 and made a deposit of $84.17. His bank service charge was $4.50. What was his balance after these transactions?

10. Scott's bank statement showed he ended the month of November with $129.40 in his checking account. All his checks had cleared. He deposited a $472.20 paycheck and wrote a rent check of $497.50. He withdrew $50.00 in cash from his checking account. How much money is left in his checking account?

11. Parker had a balance of $47.50 on January 1. He deposited $210.95 on January 5 and then paid bills. He wrote checks for $60.75 for the telephone bill, $55.60 for the electric bill, and $36.60 for the water bill. How much was in his account after paying bills?

CHAPTER 6 REVIEW

Round the following numbers to the nearest tenth.

1. 32.235 _____

2. 4.561 _____

3. 0.075 _____

Round the following numbers to the nearest hundredth.

4. 501.479 _____

5. 72.492 _____

6. 30.8472 _____

Round the following numbers to the nearest thousandth.

7. 2.0392 _____

8. 0.0057212 _____

9. 14.0851 _____

Write out in words.

10. 405.95 _____

_____

11. 5.047 _____

_____

12. 0.746 _____

_____

_____

Express in digits.

13. One hundred ninety-two and thirty-five hundredths

_____

14. Two and twenty-nine hundredths

_____

15. Seven thousand, four hundred two and three hundred sixty-three ten-thousandths

_____

Add.

16. $12.589 + 5.62 + 0.9$ _____

17. $7.8 + 10.24 + 1.903$ _____

18. $152.64 + 12.3 + 0.024$ _____

Subtract.

19. $18.547 - 9.62$ _____

20. $1.85 - 0.093$ _____

21. $45.2 - 37.9$ _____

Multiply.

22. 4.58×0.025 _____

23. 0.879×1.7 _____

24. 30.7×0.0041 _____

Divide.

25. $17.28 \div .054$ _____

26. $174.66 \div 1.23$ _____

27. $2.115 \div 9$ _____

Change to a fraction.

28. 0.55 _____

29. 0.84 _____

30. 0.32 _____

Change to a mixed number.

31. 7.375 _____

32. 9.6 _____

33. 13.25 _____

Change to a decimal.

34. $5\frac{3}{25}$ _____

35. $\frac{7}{100}$ _____

36. $10\frac{2}{3}$ _____

Put the following sets of decimals in order from GREATEST to LEAST.

37. 0.5, 0.55, 0.505, 0.05

_____

38. 0.24, 0.201, 0.022, 0.2

_____

39. 1.89, 1.08, 0.98, 0.9

_____

Put the following sets of decimals in order from LEAST to GREATEST.

40. 0.59, 0.5, 0.059, 0.509

_____

41. 0.19, 0.2, 0.109, 0.22

_____

42. 1.75, 0.79, 1.709, 1.8

_____

43. Super-X sells tires for $24.56 each. Save-Rite sells the identical tire for $21.97. How much can you save by purchasing a tire from Save-Rite?

_____

44. White rice sells for $5.64 for a 20 pound bag. A three pound bag costs $1.59. Which is the better buy?

_____

45. Daisy's Discount Mart sells 500 sheet packs of notebook paper for $4.00, 125 sheet packs for $1.17, and 200 sheet packs for $1.40. Which is the best buy?

_____

46. Xandra bought a mechanical pencil for $2.38, 3 pens for $0.89 each, and a pack of graph paper for $3.42. She paid $0.42 tax. What was her change from a ten dollar bill?

_____

47. Charlie makes $13.45 per hour repairing lawn mowers part time. If he worked 15 hours, how much was his gross pay?

_____

48. Gene works for his father sanding wooden rocking chairs. He earns $6.35 per chair. How many chairs does he need to sand in order to buy a portable radio/CD player for $146.05?

_____

49. Margo's Mint Shop has a machine that produces 4.35 pounds of mints per hour. How many pounds of mints are produced in each 8 hour shift?

_____

50. Carter's Junior High track team ran the first leg of a 400 meter relay race in 10.23 seconds, the second leg in 11.4 seconds, the third leg in 10.77 seconds, and the last leg in 9.9 seconds. How long did it take for them to complete the race?

_____

UNDERSTANDING EXPONENTS

Sometimes it is necessary to multiply a number by itself one or more times. For example, in a math problem, you may need to multiply 3×3 or $5 \times 5 \times 5 \times 5$. In these situations, mathematicians have come up with a shorter way of writing out this kind of multiplication. Instead of writing 3×3, you can write 3^2, or, instead of $5 \times 5 \times 5 \times 5$, 5^4 means the same thing. The first number is the **base**. The small, raised number is called the **exponent**. The exponent tells how many times the base should be multiplied by itself.

EXAMPLE 1: 6^3 ← exponent, ← base   **This means multiply 6 three times:** $6 \times 6 \times 6$

You also need to know two special properties of exponents:

> 1. Any base number raised to the exponent of 1 equals the base number.
> 2. Any base number raised to the exponent of 0 equals 1.

EXAMPLE 2: $4^1 = 4$ $10^1 = 10$ $25^1 = 25$
$4^0 = 1$ $10^0 = 1$ $25^0 = 1$

Rewrite the following problems using <u>exponents</u>.

EXAMPLE: $2 \times 2 \times 2 = 2^3$

1. $4 \times 4 \times 4 =$ _____
2. $5 \times 5 \times 5 \times 5 =$ _____
3. $11 \times 11 =$ _____
4. $8 \times 8 =$ _____
5. $17 \times 17 \times 17 =$ _____
6. $12 \times 12 =$ _____
7. $1 \times 1 \times 1 \times 1 \times 1 =$ _____
8. $7 \times 7 \times 7 \times 7 \times 7 \times 7 =$ _____
9. $100 \times 100 \times 100 =$ _____

Use your calculator to figure what number each number with an exponent represents.

EXAMPLE: $2^3 = 2 \times 2 \times 2 = 8$

10. $2^5 =$ _____
11. $10^2 =$ _____
12. $8^3 =$ _____
13. $3^4 =$ _____
14. $25^1 =$ _____
15. $10^5 =$ _____
16. $15^0 =$ _____
17. $9^2 =$ _____
18. $3^3 =$ _____

Express each of the following numbers as a number with an exponent.

EXAMPLE: $4 = 2 \times 2 = 2^2$

19. $32 =$ _____
20. $64 =$ _____ or _____
21. $1000 =$ _____
22. $27 =$ _____
23. $81 =$ _____ or _____
24. $121 =$ _____
25. $16 =$ _____ or _____
26. $8 =$ _____
27. $49 =$ _____

MULTIPLYING AND DIVIDING BY MULTIPLES OF TEN

Multiplying and dividing decimal numbers by multiples of ten is easy. To **multiply**, simply move the decimal point to the **right**, and to **divide**, move the decimal point to the **left**.

> **Rule for multiplying by multiples of ten:** Move the decimal point to the right the same number of spaces as zeros.

EXAMPLE 1: 5.43×1000

LONG METHOD	SHORT METHOD
5.43 $\times 1000$ 3000 4000 5000 $5430.00 = 5,430$	1. Copy the number: 5.43 2. Count the number of zeros in 1000. (3) 3. Move the decimal point to the right 3 places. You will need to add a zero. 5.430. $5.43 \times 1000 = 5,430$

> **Rule for dividing by multiples of ten:** Move the decimal point to the left the same number of spaces as zeros.

EXAMPLE 2: $7.21 \div 100$

LONG METHOD	SHORT METHOD
$\begin{array}{r} 0.0721 \\ 100\overline{)7.21} \\ 7\ 00 \\ 210 \\ 200 \\ 100 \\ 100 \\ 0 \end{array}$	1. Copy the number: 7.21 2. Count the number of zeros in 100. (2) 3. Move the decimal point to the left 2 places. You will need to add a zero. .07.21 $7.21 \div 100 = 0.0721$

Simplify the following problems.

1. $7.54 \times 100 =$ _____
2. $24.95 \div 10 =$ _____
3. $0.627 \times 1000 =$ _____
4. $935.4 \div 10 =$ _____
5. $13.4 \times 1000 =$ _____
6. $1.975 \times 100 =$ _____
7. $18.6 \div 10,000 =$ _____
8. $0.59 \div 10 =$ _____

9. $17.25 \div 100 =$ _____
10. $54.9 \times 1000 =$ _____
11. $0.3 \times 100 =$ _____
12. $7.314 \div 10 =$ _____
13. $185.1 \div 100 =$ _____
14. $1.8 \times 10,000 =$ _____
15. $26.5 \div 100 =$ _____
16. $0.625 \div 10 =$ _____

17. $8.05 \times 100 =$ _____
18. $2.9 \times 1000 =$ _____
19. $14.32 \div 10 =$ _____
20. $9.12 \times 10 =$ _____
21. $400.4 \div 100 =$ _____
22. $0.418 \div 10 =$ _____
23. $3.952 \times 100 =$ _____
24. $12.68 \times 10 =$ _____

THE METRIC SYSTEM

The metric system uses units based on multiples of ten. The basic units of measure in the metric system are the **meter**, the **liter**, and the **gram**. Metric prefixes tell what multiple of ten the basic unit is multiplied by. Below is a chart of metric prefixes and their values. The ones rarely used are shaded.

Prefix	kilo (k)	hecto (h)	deka (da)	unit (m, L, g)	deci (d)	centi (c)	milli (m)
Meaning	1000	100	10	1	0.1	0.01	0.001

Multiply when changing from a greater unit to a smaller one; **divide** when changing from a smaller unit to a larger one. **The chart is set up to help you know how far and which direction to move a decimal point when making conversions from one unit to another.**

UNDERSTANDING METERS

The basic unit of **length** in the metric system is the **meter**. Meter is abbreviated "m".

Metric Unit	Abbreviation	Memory Tip	Equivalents
1 millimeter	mm	Thickness of a dime	10 mm = 1 cm
1 centimeter	cm	Width of the tip of the little finger	100 cm = 1 m
1 meter	m	Distance from the nose to the tip of fingers (a little longer than a yard)	1000 m = 1 km
1 kilometer	km	A little more than half a mile	

UNDERSTANDING LITERS

The basic unit of **liquid volume** in the metric system is the **liter**. Liter is abbreviated "L".

The liter is the volume of a cube measuring 10 cm on each side. A milliliter is the volume of a cube measuring 1 cm on each side. A capital L is used to signify liter, so it is not confused with the number 1.

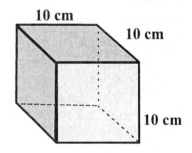

Volume = 1000 cm^3 = 1 Liter
(a little more than a quart)

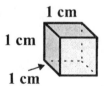

Volume = 1 cm^3 = 1 mL
(an eyedropper holds 1 mL)

UNDERSTANDING GRAMS

The basic unit of **mass** in the metric system is the **gram**. Gram is abbreviated "g".

A **gram** is the **weight** of **one cubic centimeter** of **water** at 4^0 C.

A large paper clip weighs about 1 gram (1g).
A nickel weighs 5 grams (5 g).
1000 grams = 1 kilogram (kg) = a little over 2 pounds

1 milligram (mg) = 0.001 gram. This is an extremely small amount and is used in medicine.

An aspirin tablet weighs 300 mg.

CONVERTING UNITS WITHIN THE METRIC SYSTEM

Converting units such as kilograms to grams or centimeters to decimeters is easy now that you know how to multiply and divide by multiples of ten.

Prefix	kilo (k)	hecto (h)	deka (da)	unit (m, L, g)	deci (d)	centi (c)	milli (m)
Meaning	1000	100	10	1	0.1	0.01	0.001

EXAMPLE 1: 2 L = _____ mL

$2.000 L = 2000$ mL

Look at the chart above. To move from liters to milliliters, you move to the right three places. So, to convert the 2 L to mL, move the decimal point three places to the right. You will need to add zeros.

EXAMPLE 2: 5.25 cm = _____ m

005.25 cm = 0.0525 m

To move from centimeters to meters, you need to move two spaces to the left. So, to convert 5.25 cm to m, move the decimal point two spaces to the left. Again, you need to add zeros.

Solve the following problems.

1. 35 mg = _____ g
2. 6 km = _____ m
3. 21.5 mL = _____ L
4. 4.9 mm = _____ cm
5. 5.35 kL = _____ mL
6. 32.1 mg = _____ kg
7. 156.4 m = _____ km

8. 25 mg = _____ cg
9. 17.5 L = _____ mL
10. 4.2 g = _____ kg
11. 0.06 daL = _____ dL
12. 0.417 kg = _____ cg
13. 18.2 cL = _____ L
14. 81.2 dm = _____ cm

15. 72.3 cm = _____ m
16. 0.003 kL = _____ L
17. 5.06 g = _____ mg
18. 1.058 mL = _____ cL
19. 43 hm = _____ km
20. 2.057 m = _____ cm
21. 564.3 g = _____ kg

ESTIMATING METRIC MEASUREMENTS

Choose the best estimates.

1. The height of an average man
   A. 18 cm
   B. 1.8 m
   C. 6 km
   D. 36 mm

2. The volume of a coffee cup
   A. 300 mL
   B. 20 L
   C. 5 L
   D. 1 kL

3. The width of this book
   A. 215 mm
   B. 75 cm
   C. 2 m
   D. 1.5 km

4. The weight of an average man
   A. 5 mg
   B. 15 cg
   C. 25 g
   D. 90 kg

5. The length of a basketball player's foot
   A. 2 m
   B. 1 km
   C. 30 cm
   D. 100 mm

6. The weight of a dime
   A. 3 g
   B. 30 g
   C. 10 cg
   D. 1 kg

7. The width of your hand
   A. 2 km
   B. 0.5 m
   C. 25 cm
   D. 90 mm

8. The length of a basketball court
   A. 1000 mm
   B. 250 cm
   C. 28 m
   D. 2 km

Choose the best units of measure.

9. The distance from Minneapolis to Duluth.
   A. millimeter
   B. centimeter
   C. meter
   D. kilometer

10. The length of a house key
    A. millimeter
    B. centimeter
    C. meter
    D. kilometer

11. The thickness of a nickel
    A. millimeter
    B. centimeter
    C. meter
    D. kilometer

12. The width of a classroom
    A. millimeter
    B. centimeter
    C. meter
    D. kilometer

13. The length of a piece of chalk
    A. millimeter
    B. centimeter
    C. meter
    D. kilometer

14. The height of a pine tree
    A. millimeter
    B. centimeter
    C. meter
    D. kilometer

114

SCIENTIFIC NOTATION

Mathematicians use **scientific notation** to express very large and very small numbers. **Scientific notation** expresses a number in the following form:

$$x.xx \times 10^x$$

only one digit before the decimal ➞

remaining digits not ending in zeros after the decimal ➞

multiplied by a multiple of ten

USING SCIENTIFIC NOTATION FOR LARGE NUMBERS

Scientific notation simplifies very large numbers that have many zeros. For example, Pluto averages a distance of 5,900,000,000 kilometers from the sun. In scientific notation, a decimal is inserted after the first digit (5.), the rest of the digits are copied except for the zeros at the end (5.9), and the result is multiplied by 10^9. The exponent = the total number of digits in the original number minus 1 or the number of spaces the decimal point moved.

$5,900,000,000 = 5.9 \times 10^9$ The following are more examples:

EXAMPLES: $32,560,000,000 = 3.256 \times 10^{10}$ $5,060,000 = 5.06 \times 10^6$

decimal moves 10 spaces to the left┘   └decimal moves 6 spaces to the left

Convert the following numbers to scientific notation.

1.   4,230,000,000 = _____

2.   64,300,000 = _____

3. 951,000,000,000 = _____

4.   12,300 = _____

5. 20,350,000,000 = _____

6.   9,000 = _____

7. 450,000,000,000 = _____

8.   6,200 = _____

9.   87,000,000 = _____

10.   105,000,000 = _____

11. 1,083,000,000,000 = _____

12.   304,000 = _____

To convert a number written in scientific notation back to conventional form, reverse the steps.

EXAMPLE: $4.02 \times 10^5 = 4.02000 = 402,000$ Move the decimal 5 spaces to the right and add zeros.

Convert the following numbers from scientific notation to conventional numbers.

13. 6.85×10^8 = _____

14. 1.3×10^{10} = _____

15. 4.908×10^4 = _____

16. 7.102×10^6 = _____

17. 2.5×10^3 = _____

18. 9.114×10^5 = _____

19. 5.87×10^7 = _____

20. 8.047×10^8 = _____

21. 3.81×10^5 = _____

22. 9.5×10^{12} = _____

23. 1.504×10^6 = _____

24. 7.3×10^9 = _____

USING SCIENTIFIC NOTATION FOR SMALL NUMBERS

Scientific notation also simplifies very small numbers that have many zeros. For example, the diameter of a helium atom is 0.000000000244 meters. It can be written in scientific notation as 2.44×10^{-10}. The first number is always greater than 0, and the first number is always followed by a decimal point. The negative exponent indicates how many digits the decimal point moved to the right. The exponent is negative when the original number is less than 1. To convert small numbers to scientific notation, follow the **EXAMPLES** below.

EXAMPLES: $0.00058 = 5.8 \times 10^{-4}$ $0.00003059 = 3.059 \times 10^{-5}$

decimal point moves   negative exponent   decimal moves 5 spaces to the right
4 spaces to the right   indicates the original
                             number is less than 1.

Convert the following numbers to scientific notation.

1. $0.00000254 =$ _____

2. $0.0000000508 =$ _____

3. $0.000008004 =$ _____

4. $0.00047 =$ _____

5. $0.000000005478 =$ _____

6. $0.00000059 =$ _____

7. $0.00000004712 =$ _____

8. $0.00025 =$ _____

9. $0.0000000501 =$ _____

10. $0.0000006 =$ _____

11. $0.0000000000875 =$ _____

12. $0.00004 =$ _____

Now convert small numbers written in scientific notation back to conventional form.

EXAMPLE: $3.08 \times 10^{-5} = 00003.08 = 0.0000308$ Move the decimal 5 spaces to the left and add zeros.

Convert the following numbers from scientific notation to conventional numbers.

13. $1.18 \times 10^{-7} =$ _____

14. $2.3 \times 10^{-5} =$ _____

15. $6.205 \times 10^{-9} =$ _____

16. $4.1 \times 10^{-6} =$ _____

17. $7.632 \times 10^{-4} =$ _____

18. $5.48 \times 10^{-10} =$ _____

19. $2.75 \times 10^{-8} =$ _____

20. $4.07 \times 10^{-7} =$ _____

21. $5.2 \times 10^{-3} =$ _____

22. $7.01 \times 10^{-6} =$ _____

23. $4.4 \times 10^{-5} =$ _____

24. $3.43 \times 10^{-2} =$ _____

CHAPTER 7 REVIEW

Rewrite the following problems using exponents.

1. $3 \times 3 \times 3 \times 3$ _____

2. $5 \times 5 \times 5$ _____

3. $10 \times 10 \times 10 \times 10 \times 10$ _____

4. 25×25 _____

Use a calculator to figure the solution to the following.

5. $2^2 =$ _____   8. $15^0 =$ _____

6. $5^3 =$ _____   9. $10^4 =$ _____

7. $12^1 =$ _____   10. $7^2 =$ _____

Simplify the following problems.

11. 5.623×10 = _____

12. $245.6 \div 100$ = _____

13. 14.95×1000 = _____

14. $0.365 \times 10,000$ = _____

15. $0.0587 \div 1000$ = _____

16. $1.2 \div 100$ = _____

Solve the following problems.

17. 120 m = _____ km

18. 9 g = _____ mg

19. 0.02 kL = _____ L

20. 1.5 mg = _____ g

21. 15 cm = _____ mm

22. 5 L = _____ mL

23. 0.005 kg = _____ g

24. 55 mL = _____ L

25. 30 cm = _____ m

Convert the following numbers to scientific notation.

26. $22,300,000 =$ _____

27. $5,340,000 =$ _____

28. $0.00000005874 =$ _____

29. $1,451 =$ _____

30. $0.0000041 =$ _____

31. $0.0004178 =$ _____

32. $105,000 =$ _____

33. $705,000,000 =$ _____

34. $0.0000747 =$ _____

35. $0.08 =$ _____

36. $105 =$ _____

37. $0.0048754 =$ _____

38. $62,400 =$ _____

Convert the following numbers from scientific notation to conventional numbers.

39. $5.204 \times 10^{-5} =$ _____

40. $1.02 \times 10^{7} =$ _____

41. $8.1 \times 10^{5} =$ _____

42. $2.0078 \times 10^{-4} =$ _____

43. $4.7 \times 10^{-3} =$ _____

44. $7.75 \times 10^{-8} =$ _____

45. $9.795 \times 10^{9} =$ _____

46. $3.51 \times 10^{2} =$ _____

47. $6.32514 \times 10^{3} =$ _____

48. $1.584 \times 10^{-6} =$ _____

49. $7.041 \times 10^{4} =$ _____

50. $4.09 \times 10^{-7} =$ _____

Chapter 8

PERCENTS

CHANGING PERCENTS TO DECIMALS
AND DECIMALS TO PERCENTS

Change the following percents to **decimal** numbers.

Directions: Move the **decimal** point two places to the left, and drop the **percent** sign. If there is no decimal point written, it is after the number and before the percent sign. Sometimes you will need to add a "0". (See 5% below.)

EXAMPLES: 14% = 0.14   5% = 0.05   100% = 1   103% = 1.03
(decimal point)

Change the following percents to decimal numbers.

1. 18% = _____	8. 119% = _____	15. 5% = _____
2. 23% = _____	9. 2% = _____	16. 25% = _____
3. 9% = _____	10. 55% = _____	17. 410% = _____
4. 63% = _____	11. 80% = _____	18. 1% = _____
5. 4% = _____	12. 17% = _____	19. 50% = _____
6. 45% = _____	13. 66% = _____	20. 99% = _____
7. 2% = _____	14. 13% = _____	21. 107% = _____

Change the following decimal numbers to percents.

Directions: To change a **decimal** number to a **percent**, move the **decimal** point two places to the right, and add a **percent** sign. You may need to add a "0". (See 0.8 below.)

EXAMPLES: 0.62 = 62%   0.07 = 7%   0.8 = 80%   0.166 = 16.6%   1.54 = 154%

Change the following decimal numbers to percents.

22. 0.15 = _____	29. 0.044 = _____	36. 0.042 = _____
23. 0.62 = _____	30 0.58 = _____	37. 0.375 = _____
24. 1.53 = _____	31. 0.86 = _____	38. 5.09 = _____
25. 0.22 = _____	32. 0.29 = _____	39. 0.75 = _____
26. 0.35 = _____	33. 0.06 = _____	40. 0.3 = _____
27. 0.375 = _____	34. 0.48 = _____	41. 2.9 = _____
28. 0.648 = _____	35. 3.089 = _____	42. 0.06 = _____

CHANGING PERCENTS TO FRACTIONS
AND FRACTIONS TO PERCENTS

EXAMPLE: Change 15% to a fraction.

Step 1: Copy the number without the percent sign. **15** is the top number of the fraction.

Step 2: The bottom number of the fraction is 100.

$$15\% = \frac{15}{100}$$

Step 3: Reduce the fraction. $\frac{15}{100} = \frac{3}{20}$

Change the following percents to fractions and reduce.

1.	50%	6.	63%	11.	25%	16.	40%
2.	13%	7.	75%	12.	5%	17.	99%
3.	22%	8.	91%	13.	16%	18.	30%
4.	95%	9.	18%	14.	1%	19.	15%
5.	52%	10.	3%	15.	79%	20.	84%

EXAMPLE: Change $\frac{7}{8}$ to a percent.

Step 1: Divide 7 by 8. Add as many 0's as necessary.

$$\begin{array}{r} .875 \\ 8\overline{)7.000} \\ -\underline{64} \\ 60 \\ -\underline{56} \\ 40 \\ -\underline{40} \\ 0 \end{array}$$

Step 2: Change the decimal answer, .875, to a percent by moving the decimal point 2 places to the right.

$$\frac{7}{8} = .875 = 87.5\%$$

Change the following fractions to percents.

1.	$\frac{1}{5}$	4.	$\frac{3}{8}$	7.	$\frac{1}{10}$	10.	$\frac{3}{4}$	13.	$\frac{1}{16}$	16.	$\frac{3}{4}$
2.	$\frac{5}{8}$	5.	$\frac{3}{16}$	8.	$\frac{4}{5}$	11.	$\frac{1}{8}$	14.	$\frac{1}{4}$	17.	$\frac{2}{5}$
3.	$\frac{7}{16}$	6.	$\frac{19}{100}$	9.	$\frac{15}{16}$	12.	$\frac{5}{16}$	15.	$\frac{4}{100}$	18.	$\frac{16}{25}$

119

CHANGING PERCENTS TO MIXED NUMBERS
AND MIXED NUMBERS TO PERCENTS

EXAMPLE: Change 218% to a fraction.

Step 1: Copy the number without the percent sign. **218** is the top number of the fraction.

Step 2: The bottom number of the fraction is **100**.

$$218\% = \frac{218}{100}$$

Step 3: Reduce the fraction and convert to a mixed number. $\frac{218}{100} = \frac{109}{50} = 2\frac{9}{50}$

Change the following percents to mixed numbers.

1.	150%	6.	163%	11.	205%	16.	340%
2.	113%	7.	275%	12.	405%	17.	199%
3.	222%	8.	191%	13.	516%	18.	300%
4.	395%	9.	108%	14.	161%	19.	125%
5.	252%	10.	453%	15.	179%	20.	384%

EXAMPLE: Change $5\frac{3}{8}$ to a percent.

Step 1: Divide 3 by 8. Add as many 0's as necessary.

$$
\begin{array}{r}
.375 \\
8\overline{)3.000} \\
-\underline{24} \\
60 \\
-\underline{56} \\
40 \\
-\underline{40} \\
0
\end{array}
$$

Step 2: So, $5\frac{3}{8} = 5.375$. Change the decimal answer to a percent by moving the decimal point 2 places to the right.

$$5\frac{3}{8} = 5.375 = 537.5\%$$

Change the following mixed numbers to percents.

1. $5\frac{1}{2}$	4. $3\frac{1}{4}$	7. $1\frac{3}{10}$	10. $2\frac{13}{25}$	13. $1\frac{3}{16}$	16. $4\frac{4}{5}$
2. $8\frac{3}{4}$	5. $4\frac{7}{8}$	8. $6\frac{1}{5}$	11. $1\frac{1}{8}$	14. $1\frac{1}{16}$	17. $3\frac{2}{5}$
3. $1\frac{5}{8}$	6. $2\frac{3}{100}$	9. $4\frac{7}{10}$	12. $2\frac{5}{16}$	15. $5\frac{17}{100}$	18. $2\frac{17}{100}$

120

COMPARING THE RELATIVE MAGNITUDE OF NUMBERS

When comparing the relative magnitude of numbers, the greater than (>), less than (<), and the equal to (=) signs are the ones most frequently used. The simplest way to compare numbers that are in different notations, like percent, decimals, and fractions, is to change all of them to one notation. Decimals are the easiest to compare.

EXAMPLE 1: Which is larger: $1\frac{1}{4}$ or 1.3?

Answer: Change $1\frac{1}{4}$ to a decimal. $\frac{1}{4} = .25$, so $1\frac{1}{4} = 1.25$.
1.3 is larger than 1.25, so $1.3 > 1\frac{1}{4}$.

EXAMPLE 2: Which is smaller: 60% or $\frac{2}{3}$?

Answer: Change both to decimals.
$60\% = .6$ and $\frac{2}{3} = .\overline{66}$
.6 is smaller than $.\overline{66}$, so $60\% < \frac{2}{3}$

Fill in each box with the correct sign.

1. 23.4 ☐ $23\frac{1}{2}$

2. 17% ☐ $.17$

3. $\frac{3}{8}$ ☐ 37.5%

4. 25% ☐ $\frac{2}{10}$

5. 234% ☐ 23.4

6. $\frac{1}{7}$ ☐ 14%

7. 13.95 ☐ $13\frac{8}{9}$

8. 4.0 ☐ 40%

9. 25% ☐ $\frac{3}{2}$

10. $\frac{12}{4}$ ☐ 300%

3. 6% ☐ $\frac{1}{16}$

6. 1.33 ☐ $\frac{4}{3}$

13. $.8$ ☐ $\frac{4}{5}$

14. 75% ☐ $\frac{3}{4}$

15. $\frac{5}{8}$ ☐ 62%

Compare the sums, differences, products, and quotients below. Fill in each box with the correct sign.

1. $(32+15)$ ☐ $(65-17)$

2. $(45-13)$ ☐ $(31+9)$

3. $(24\div4)$ ☐ $(24\div6)$

4. $(48\div6)$ ☐ (4×3)

5. (4×3) ☐ $(48\div6)$

6. (18×4) ☐ (5×17)

7. $[(1+3)+5]$ ☐ $[(5+(3+1)]$

8. $[(1+(3+5)]$ ☐ $[(5-3)+1]$

9. $(25\div5)$ ☐ (5×5)

REPRESENTING RATIONAL NUMBERS GRAPHICALLY

You now know how to convert fractions to decimals, decimals to fractions, fractions to percentages, percentages to fractions, decimals to percentages, and percentages to decimals. Study the examples below to understand how fractions, decimals, and percentages can be expressed graphically.

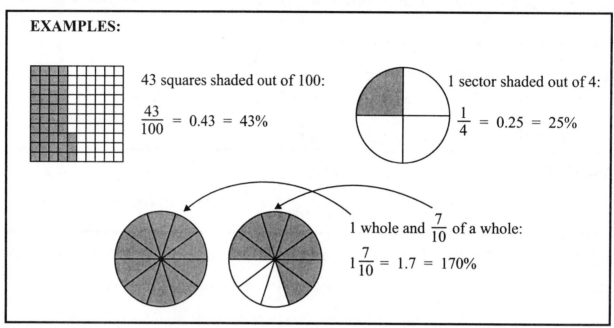

EXAMPLES:

43 squares shaded out of 100:

$$\frac{43}{100} = 0.43 = 43\%$$

1 sector shaded out of 4:

$$\frac{1}{4} = 0.25 = 25\%$$

1 whole and $\frac{7}{10}$ of a whole:

$$1\frac{7}{10} = 1.7 = 170\%$$

Fill in the missing information in the chart below. Shade in the graphic for the problems that are not done for you. Reduce all fractions to lowest terms.

Graphic	Fraction	Decimal	Percent	Graphic	Fraction	Decimal	Percent
1.				5.			125%
2.		0.92		6.			
3.				7.			
4.	$\frac{4}{5}$			8.			

FINDING THE PERCENT OF THE TOTAL

EXAMPLE: There were 75 customers at Bill's gas station this morning. Thirty-two percent used a credit card to make their purchase. How many customers used credit cards this morning at Bill's?

Step 1: Change 32% to a decimal. .32

Step 2: Multiply by the total number mentioned.

$$\begin{array}{r} .32 \\ \times\, 75 \\ \hline 160 \\ 224 \\ \hline 24.00 \end{array}$$

24 customers used credit cards.

Find the percent of the total in the problems below.

1. Eighty-five percent of Mrs. Coomer's math class passed her final exam. There were 40 students in her class. How many passed?

2. Fifteen percent of a bag of chocolate candies have a red coating on them. How many red pieces are in a bag of 60 candies?

3. Sixty-eight percent of Valley Creek School students attended this year's homecoming dance. There are 675 students. How many attended the dance?

4. Out of the 4,500 people who attended the rock concert, forty-six percent purchased a T-shirt. How many people bought T-shirts?

5. Nina sold ninety-five percent of her 500 cookies at the bake sale. How many cookies did she sell?

6. Twelve percent of yesterday's customers purchased premium grade gasoline from GasCo. If GasCo had 200 customers, how many purchased premium grade gasoline?

7. The Candy Shack sold 138 pounds of candy on Tuesday. Fifty-two percent of the candy was jelly beans. How many pounds of jelly beans were sold Tuesday?

8. A fund-raiser at the school raised $617.50. Ninety-four percent went to local charities. How much money went to charities?

9. Out of the company's $6.5 million profit, eight percent will be paid to shareholders as dividends. How much will be paid out in dividends?

10. Ted's Toys sold seventy-five percent of its stock of stuffed bean animals on Saturday. If Ted's Toys had 620 originally in stock, how many were sold on Saturday?

TIPS AND COMMISSIONS

Vocabulary

Tip: A **tip** is money given to someone doing a service for you such as a waiter, waitress, porter, cab driver, beautician, etc.

Commission: In many businesses, sales people are paid on **commission** - a percent of the total sales they make.

Problems requiring you to figure a tip, commission, or percent of a total are all done in the same way.

EXAMPLE: Ramon made a 4% commission on an $8,000 pickup truck he sold. How much was his commission?

$$
\begin{array}{rr}
\textbf{TOTAL COST} & \textbf{\$8,000} \\
\times \underline{\textbf{RATE OF COMMISSION}} & \times\ \underline{\textbf{0.04}} \\
\textbf{COMMISSION} & \textbf{\$320.00}
\end{array}
$$

Solve each of the following problems.

1. Whitney makes 12% commission on all her sales. This month she sold $9,000 worth of merchandise. What was her commission?

_____

2. Marcus gives 25% of his income to his parents to help cover expenses. He earns $340 per week. How much money does he give his parents?

_____

3. Jan pays $640 per month for rent. If the rate of inflation is 5%, how much can Jan expect to pay monthly next year?

_____

4. The total bill at Jake's Catfish Place came to $35.80. Jim wanted to leave a 15% tip. How much money will he leave for the tip?

_____

5. Rami makes $2,400 per month and puts 6% in a savings plan. How much does he save per month?

_____

6. Cristina makes $2,550 per month. Her boss promised her a 7% raise. How much more will she make per month?

_____

7. Out of 150 math students, 86% passed. How many students passed math class?

_____

8. Marta sells Sue Ann Cosmetics and gets 20% commission on all her sales. Last month, she sold $560.00 worth of cosmetics. How much was her commission?

_____

FINDING THE AMOUNT OF DISCOUNT

Sale prices are sometimes marked 30% off, or better yet, 50% off. A 30% DISCOUNT means you will pay 30% less than the original price. How much money you save is also known as the amount of the discount. Read the **EXAMPLE** below to learn to figure the amount of a discount.

EXAMPLE: A $179.00 chair is on sale for 30% off . How much can I save if I buy it now?

Step 1:   Change 30% to a decimal.   30% = .30

Step 2:   Multiply the original price by the discount.

ORIGINAL PRICE	$179.00
× **% DISCOUNT**	× .30
SAVINGS	$ 53.70

Practice finding the amount of the discount. Round off answers to the nearest penny.

1. Tubby Tires is offering a 25% discount on tires purchased on Tuesday. How much can you save if you buy tires on Tuesday regularly priced at $225.00 any other day of the week? _____

2. The regular price for a garden rake is $10.97 at Sly's Super Store. This week, Sly is offering a 30% discount. How much is the discount on the rake? _____

3. Christine bought a sweater regularly priced at $26.80 with a coupon for 20% off any sweater. How much did she save? _____

4. The software that Marge needs for her computer is priced at $69.85. If she waits until a store offers it at 20% off, how much will she save? _____

5. Ty purchased jeans that were priced $23.97. He received a 15% employee discount. How much did he save? _____

6. The Bakery Company offers a 60% discount on all bread made the day before. How much can you save on a $2.40 loaf made today if you wait until tomorrow to buy it? _____

7. A furniture store advertises a 40% off liquidation sale on all items. How much would the discount be on a $2,530 dining room set? _____

8. Becky bought a $4.00 nail polish on sale for 30% off. What was the dollar amount of the discount? _____

9. How much is the discount on a $350 racing bike marked 15% off? _____

10. Raymond receives a 2% discount from his credit card company on all purchases made with the credit card. What is his discount on $1,575.50 worth of purchases? _____

FINDING THE DISCOUNTED SALE PRICE

To find the discounted sale price, you must go one step further than shown on the previous page. Read the example below to learn how to figure **discount** prices.

EXAMPLE: A $74.00 chair is on sale for 25% off. How much can I save if I buy it now?

Step 1: Change 25% to a decimal.   25% = .25

Step 2: Multiply the original price by the discount.

ORIGINAL PRICE	$74.00
× % DISCOUNT	**× .25**
SAVINGS	$18.50

Step 3: Subtract the savings amount from the original price to find the sale price.

ORIGINAL PRICE	$74.00
− SAVINGS	**− 18.50**
SALE PRICE	$55.50

Figure the sale price of the items below. The first one is done for you.

ITEM	PRICE	% OFF	MULTIPLY	SUBTRACT	SALE PRICE
1. pen	$1.50	20%	1.50 × .2 = $0.30	1.50 − 0.30 = 1.20	$1.20
2. recliner	$325	25%			
3. juicer	$55	15%			
4. blanket	$14	10%			
5. earrings	$2.40	20%			
6. figurine	$8	15%			
7. boots	$159	35%			
8. calculator	$80	30%			
9. candle	$6.20	50%			
10. camera	$445	20%			
11. VCR	$235	25%			
12. video game	$25	10%			

SALES TAX

EXAMPLE: The total price of a sofa is $560.00 + 6% sales tax. How much is the sales tax? What is the total cost?

Step 1: You will need to change 6% to a decimal. 6% = .06

Step 2: Simply multiply the cost, $560, by the tax rate, 6%. 560 × .06 = 33.6
The answer will be $33.60. (You need to add a 0 to the answer. When dealing with money, there needs to be two places after the decimal point).

COST	$560
× 6% TAX	× .06
SALES TAX	$33.60

Step 3: Add the sales tax amount, $33.60 to the cost of the item sold, $560. This is the total cost.

COST	$560.00
SALES TAX	+ 33.60
TOTAL COST	$593.60

NOTE: When the answer to the question involves money, you always need to round off the answer to the nearest hundredth (2 places after the decimal point). Sometimes you will need to add a zero.

Figure the total costs in the problems below. The first one is done for you.

ITEM	PRICE	% SALES TAX	MULTIPLY	ADD PRICE PLUS TAX	TOTAL
1. jeans	$42	7%	$42 × 0.07 = $2.94	42 + 2.94 = 44.94	$44.94
2. truck	$17,495	6%			
3. film	$5.89	8%			
4. T-shirt	$12	5%			
5. football	$36.40	4%			
6. soda	$1.78	5%			
7. 4 tires	$105.80	10%			
8. clock	$18	6%			
9. burger	$2.34	5%			
10. software	$89.95	8%			

FINDING THE PERCENT

EXAMPLE: 15 is what percent of 60?

Step 1: To solve these problems, simply divide the smaller amount by the larger amount. You will need to add a decimal point and two 0's.

$$
\begin{array}{r}
.25 \\
60\overline{)15.00} \\
-12\ 0 \\
\hline
3\ 00 \\
-3\ 00 \\
\hline
0
\end{array}
$$

Step 2: Change the answer, .25, to a percent by moving the decimal point two places to the right.

.25 = 25%   15 is 25% of 60.

Remember: To change a decimal to a percent, you will sometimes have to add a zero when moving the decimal point two places to the right.

Find the following percents.

1. What percent of 50 is 16?

2. 20 is what percent of 80?

3. 9 is what percent of 100?

4. 19 is what percent of 95?

5. Ruth made 200 cookies for the picnic. Only 25 were left at the end of the day. What percent of the cookies were left?

6. Pat made 116 bird houses to sell at the county fair. The first day he sold 29. What percent of the bird houses sold?

7. Eileen planted 90 sweet corn seeds but only 18 plants came up. What percent of the seeds germinated?

8. Tomika invests $36 of her $240 paycheck in a retirement account. What percent of her pay is she investing?

9. Ray sold a house for $115,000, and his commission was $9,200. What percent commission did he make?

10. Peter was making $16.00 per hour. After one year, he received a $2.00 per hour raise. What percent raise did he get?

11. Calvin budgets $235 per month for food. If his salary is $940 per month, what percent of his salary does he budget for food?

12. Katie earned $45 on commission for her sales totaling $225. What percent was her commission?

UNDERSTANDING SIMPLE INTEREST *

I = PRT is a formula to figure out the **cost of borrowing money** or the **amount you earn** when you **put money in a savings account**. When you want to buy a used truck or car, you go to the bank to borrow the $7,000 you need. The bank will charge you interest on the $7,000. If the simple interest rate is 9% for four years, you can figure the cost of the interest with this formula.

First, you need to understand these terms:

> **I** = Interest = The amount charged by the bank or other lender
> **P** = Principal = The amount you borrow
> **R** = Rate = The interest rate the bank is charging you
> **T** = Time = How many years you will take to pay off the loan

EXAMPLE:

> In the problem above: **I = PRT**   This means the **interest** equals the **principal** times the **rate** times the **time** in **years.**

$$I = \$7,000 \times 9\% \times 4 \text{ years}$$
$$I = \$7,000 \times .09 \times 4$$
$$I = \$2,520$$

Use the formula I = PRT to work the following problems:

1. Craig borrowed $1,800 from his parents to buy a stereo. His parents charged him 3% simple interest for 2 years. How much interest did he pay his parents? _____

2. Raul invested $5,000 in a savings account that earned 2% simple interest. If he kept the money in the account for 5 years, how much interest did he earn? _____

3. Bridgette borrowed $11,000 to buy a car. The bank charged 12% simple interest for 7 years. How much interest did she pay the bank? _____

4. A tax accountant invested $25,000 in a money market account for 3 years. The account earned 5% simple interest. How much interest did the accountant make on his investment? _____

5. Linda Kay started a savings account for her nephew with $2,000. The account earned 6% simple interest. How much interest did the account accumulate in 3 years? _____

6. Renada bought a living room set on credit. The set sold for $2,300 and the store charged her 9% simple interest for one year. How much interest did she pay? _____

7. Duane took out a $3,500 loan at 8% simple interest for 3 years. How much interest did he pay for borrowing the $3,500? _____

* Simple interest is not commonly used by banks and other lending institutions. Compound interest is more commonly used, but its calculations are more complicated and beyond the scope of the material presented in this text.

BUYING ON CREDIT

Many stores will allow you to choose between paying cash or buying on credit. When you buy on credit, you make a down payment and monthly payments until the item is paid for. You will always pay more for an item by buying on credit.

EXAMPLE: Darlene saw a living room set advertised for $899.00 or $100.00 down and $60.00 per month for 18 months. How much can she save by paying cash?

Step 1: Multiply the number of months by the monthly payments. $60.00 \times 18 = $1080

Step 2: Add the down payment to the total cost of the payments. $1,080 + $100 = $1,180

Step 3: Subtract the cash price from the total cost of paying on credit. $1,180 - $899 = $281

Answer: Darlene can save $281 by paying cash for the living room set.

For each problem below, determine how much you can save if you pay cash.

1.

Notebook Computer
$2,595 cash
or
$300 down and $75 per mo.
for 36 mo.

3.

Leather Sofa

$1,500 cash
or
$100 down and
$68 per mo. for 24 mo.

2.

SALE
Big Al's Used Jeeps

$6,000 cash
OR
only $500 down &
$250 per mo. for 30 months

4.

JET SKI SALE

$3,395 cash
or finance for
$250 down &
$145 per mo. for 24 mo.

CHAPTER 8 REVIEW

Change the following percents to decimals.

1. 45% _____
2. 219% _____
3. 22% _____
4. 1.25% _____

Change the following decimals to percents.

5. 0.52 _____
6. 0.64 _____
7. 1.09 _____
8. 0.625 _____

Change the following percents to fractions.

9. 25% _____
10. 3% _____
11. 68% _____
12. 102% _____

Change the following fractions to percents.

13. $\frac{9}{10}$ _____
14. $\frac{5}{16}$ _____
15. $\frac{1}{8}$ _____
16. $\frac{1}{4}$ _____

17. What is 1.65 written as a percent?

_____

18. What is $2\frac{1}{4}$ written as a percent?

_____

19. Change 5.65 to a percent.

_____

Fill in the equivalent numbers represented by the shaded area.

20. fraction _____

21. decimal _____

22. percent _____

Fill in the equivalent numbers represented by the shaded area.

23. fraction _____

24. decimal _____

25. percent _____

26. Uncle Howard left his only niece 56% of his assets according to his will. If his assets totaled $564,000 when he died, how much did his niece inherit?

_____

27. Celeste makes 6% commission on her sales. If her sales for a week total $4,580, what is her commission?

_____

28. Peeler's Jewelry is offering a 30% off sale on all bracelets. How much will you save if you buy a $45.00 bracelet during the sale?

_____

29. How much would an employee pay for a $724.00 stereo if the employee got a 15% discount?

_____

30. Misha bought a CD for $14.95. If sales tax was 7%, how much did she pay total?

_____

31. The Pep band made $640 during a fund-raiser. The band spent $400 of the money on new uniforms. What percent of the total did they spend on uniforms?

_____

32. Linda took out a simple interest loan for $7,000 at 11% interest for 5 years. How much interest did she have to pay back?

_____

33. McMartin's is offering a deal on fitness club memberships. You can pay $999 up front for a 3 year membership, or pay $200 down and $30 per month for 36 months. How much would you save by paying up front?

_____

34. Patton, Patton, and Clark, a law firm, won a malpractice law suit for $4,500,000. Sixty-eight percent went to the law firm. How much did the law firm make?

_____

35. Jeneane earned $340.20 commission by selling $5,670 worth of products. What percent commission did she earn?

_____

36. Tara put $500 in a savings account that earned 3% simple interest. How much interest did she make after 5 years?

_____

37. Clay bought a pair of basketball shoes for $79.99 plus 5% sales tax. What was his total cost for the shoes?

_____

38. Marsha earns a 12% commission on all jewelry sales. How much commission will she make if she sells a $564.00 necklace?

_____

39. The following advertisement is in the newspaper:

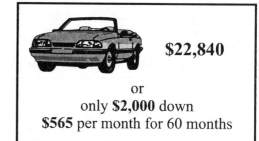

$22,840

or
only **$2,000** down
$565 per month for 60 months

How much more would you pay to buy the car on credit?

_____

40. A department store is selling all swimsuits for 40% off in August. How much would you pay for a swimsuit that is normally priced at $35.80?

_____

41. How much can Jane save by waiting to buy a $34.00 sweater that will go on sale for 20% off on Saturday?

_____

132

PROBLEM-SOLVING
AND CRITICAL THINKING

MISSING INFORMATION

Problems can only be solved if you are given enough information. Sometimes you are not given enough information to solve a problem.

EXAMPLE: Chuck has worked on his job for 1 year now. At the end of a year, his employer gave him a 12% raise. How much does Chuck make now?

To solve this problem, you need to know how much Chuck made when he began his job one year ago.

Each problem below does not give enough information for you to solve it. Beneath each problem, describe the information you would need to solve the problem.

1.   Fourteen percent of the coated chocolate candies in Nate's bag were yellow. At that rate, how many of the candies were yellow?

_____

2.   Patrick is putting up a fence around all four sides of his back yard. The fence costs $2.25 per foot, and his yard is 150 feet wide. How much will the fence cost?

_____

3.   Staci worked 5 days last week. She made $6.75 per hour before taxes. What was her total earnings before taxes were taken out?

_____

4.   Which is a better buy: 4 oz bar of soap for 88¢ or a bath bar for $1.20?

_____

5.   Randy bought a used car for $4,568 plus sales tax. What was the total cost of the car?

_____

6.   The Hall family ate at a restaurant, and each of their dinners cost $5.95. They left a 15% tip. What was the total amount of the tip?

_____

7.   If a kudzu plant grows 3 feet per day, in what month will it be 90 feet long?

_____

8.   Bethany traveled by car to her sister's house in White Bear Lake. She traveled at an average speed of 52 miles per hour. She arrived at 4:00 p.m. How far did she travel?

_____

9. Terrence earns $7.50 per hour plus 5% commission on total sales over $500 per day. Today he sold $6,500 worth of merchandise. How much did he earn for the day?

_____

10. Michelle works at a department store and gets an employee's discount on all of her purchases. She wants to buy a sweater that sells for $38.00. How much will the sweater cost after her discount?

_____

11. John filled his car with 10 gallons of gas and paid for the gas with a $20 bill. How much change did he get back?

_____

12. Olivia budgets $5.00 per work day for lunch. How much does she budget for lunch each month?

_____

13. Joey worked 40 hours and was paid $356.00. His friend Pete worked 38 hours. Who was paid more per hour?

_____

14. A train trip from Columbia to Boston took $18\frac{1}{4}$ hours. How many miles apart are the two cities?

_____

15. Caleb spent 35% of his check on rent, 10% on groceries, and 18% on utilities. How much money did he have left from his check?

_____

16. The Lyons family spent $54.00 per day plus tax on lodging during their vacation. How much tax did they pay for lodging per day?

_____

17. Richard bought cologne at a 30% off sale. How much did he save buying the cologne on sale?

_____

18. The bottling machine works 7 days a week and fills 1,000 bottles per hour. How many bottles did it fill last week?

_____

19. Tyler, who works strictly on commission, brought in $25,000 worth of sales in the last 10 days. How much was his commission?

_____

20. Ninety percent of the student body at Parks Middle School bought raffle tickets to help the basketball team buy new uniforms. The main prize was a 25-inch color TV with built-in VCR. How many students bought raffle tickets?

_____

EXACT INFORMATION

Most word problems supply exact information and ask for exact answers. The following problems are the same as those on the previous two pages with the missing information given. **Find the exact solution.**

1. Fourteen percent of the coated chocolate candies in Nate's bag were yellow. If there were 50 pieces in the bag, how many of the candies were yellow?

   _____

2. Patrick is putting up a fence around all four sides of his back yard. The fence costs $2.25 per foot. His yard is 150 feet wide and 200 feet long. How much will the fence cost?

   _____

3. Staci worked 5 days last week, 8 hours each day. She made $6.75 per hour before taxes. How much did she make last week before taxes were taken out?

   _____

4. Which is a better buy: 4 oz bar of soap for 88¢ or a 6 oz bath bar for $1.20?

   _____

5. Randy bought a used car for $4,568 plus 6% sales tax. What was the total cost of the car?

   _____

6. The Hall family ate at a restaurant, and each of the 4 dinners cost $5.95. They left a 15% tip. What was the total amount of the tip?

   _____

7. Kudzu is a rapid-growing vine found in southeastern states of the U.S. If a kudzu plant grows 3 feet per day, in what month will it be 90 feet long if it takes root in the middle of May?

   _____

8. Bethany traveled from Fairmont by car to her sister's house in Moorhead. She traveled at an average speed of 52 miles per hour. She left at 10:00 a.m. and arrived at 4:00 p.m. How far did she travel?

   _____

9. Terrence earns $7.50 per hour plus 5% commission on total sales over $500 per day. Today he sold $6,500 worth of merchandise and worked 7 hours. How much did he earn for the day?

   _____

10. Michelle works at a department store and gets a 20% employee's discount on all of her purchases. She wants to buy a sweater that sells for $38.00. How much will the sweater cost after her discount?

    _____

11. John filled his car with 10 gallons of gas priced at $1.24 per gallon. He paid for the gas with a $20 bill. How much change did he get back?

_____

12. Olivia budgets $5.00 per work day for lunch. How much does she budget for lunches if she works 21 days this month?

_____

13. Joey worked 40 hours and was paid $356.00. His friend Pete worked 38 hours at $8.70 per hour. Who was paid more per hour?

_____

14. A train trip from Columbia, SC to Boston, MA took $18\frac{1}{4}$ hours. How many miles apart are the two cities if the train travels at an average speed of 50 miles per hour?

_____

15. Caleb spent 35% of his check on rent, 10% on groceries, and 18% on utilities. How much money did he have left from his $260 check?

_____

16. The Lyons family spent $54.00 per day plus 10% tax on lodging during their vacation. How much tax did they pay per day?

_____

17. Richard bought cologne at a 30% off sale. The cologne was regularly priced at $44. How much did he save buying the cologne on sale?

_____

18. The bottling machine works 7 days a week, 14 hours per day and fills 1,000 bottles per hour. How many bottles did it fill last week?

_____

19. Tyler, who works strictly on commission, brought in $25,000 worth of sales in the last 10 days. He earns 15% commission on his sales. How much was his commission?

_____

20. Ninety percent of the total 540 students at Parks High School bought raffle tickets to help the basketball team buy new uniforms. How many students bought raffle tickets?

_____

EXTRA INFORMATION

In each of the following problems, there is extra information given. **Look closely at the question,** and use only the information you need to answer it.

EXAMPLE: Gary was making $6.50 per hour. His boss gave him a 52¢ per hour raise. Gary works 40 hours per week. What percent raise did Gary receive?

Solution: To figure the percent of Gary's raise, you do **not** need to know how many hours per week Gary works. That is extra information not needed to answer the question. To figure the percent increase, simply divide the change in pay, $0.52, by the original wages, $6.50. $0.52 \div 6.50 = 0.08$. Move the decimal two places to the right to determine the percent of the raise.
Gary received an 8% raise.

In the following questions, determine what information is needed from the problem to answer the question and solve.

1. Leah wants a new sound system that is on sale for 15% off the regular price of $420. She has already saved $325 toward the cost. What is the dollar amount of the discount?

_____

2. Praveen bought a shirt for $34.80 and socks for $11.25. He gets $10.00 per week for his allowance. He paid $2.76 sales tax. What was his change from three $20 bills?

_____

3. Marty worked 38 hours this week, and he earned $8.40 per hour. His taxes and insurance deductions amount to 34% of his gross pay. What is his total gross pay?

_____

4. Tamika went shopping and spent $4.80 for lunch. She wants to buy a sweater that is on sale for $\frac{1}{4}$ off the regular $56.00 price. How much will she save?

_____

5. Nick drove an average of 52 miles per hour for 7 hours. His car gets 32 miles per gallon. How far did he travel?

_____

6. The odometer on Melody's car read 45,920 at the beginning of her trip and 46,460 at the end of her trip. Her speed averaged 54 miles per hour, and she used 20 gallons of gasoline. How many miles per gallon did she average?

_____

7. Eighty percent of the eighth graders attended the end of the year class picnic. There are 160 eighth graders and 54% of them ride the bus to school each day. How many students went to the class picnic?

_____

8. Matt has $5.00 to spend on snacks. Tastee Potato Chips cost $2.57 for a one pound bag at the grocery store. T-Mart sells the same bag of chips for $1.98. How much can he save if he buys the chips at T-Mart?

_____

9. Elaina wanted to make 10 cakes for the band bake sale. She needed $1\frac{3}{4}$ cups of flour and $2\frac{1}{4}$ cups of sugar for each cake. How many cups of flour did she need in all?

_____

ESTIMATED SOLUTIONS

Some problems require an estimated solution. In order to have enough product to complete the job, you often need to buy more than you actually need. **In the following problems, be sure to round up your answer to the next whole number to find the correct solution.**

1. Endicott Publishing received an order for 550 books. Each shipping box holds 30 books. How many boxes do the packers need to ship the order?

    _____

2. Elena's 250 chickens laid 314 eggs in the last 2 days. How many egg cartons holding one dozen eggs would be needed to hold all the eggs?

    _____

3. Antoinetta's Italian restaurant uses $1\frac{1}{4}$ quarts of olive oil every day. The restaurant is open 7 days a week. For the month of September, how many gallons should the cooks order to have enough?

    _____

4. Eastmont High School is taking 316 students and 22 chaperones on a field trip. Each bus holds 44 persons. How many buses will the school need?

    _____

5. Fran volunteered to hem 11 choir robes that came in too long. Each robe is 7 feet around at the bottom. Hemming tape comes three yards to a pack. How many packs will Fran need to buy to go around all the robes?

    _____

6. Tonya is making matching vests for the children's choir. Each vest has 5 buttons on it, and there are 23 children in the choir. The button she picked comes 6 buttons to a card. How many cards of buttons does she need?

    _____

7. Tiffany is making the bread for the banquet. She needs to make 6 batches with $2\frac{1}{4}$ lb of flour in each batch. How many 10-lb bags of flour will she need to buy?

    _____

8. The homeless shelter is distributing 250 sandwiches per day to hungry guests. It takes one foot of plastic wrap to wrap each sandwich. There are 150 feet of plastic wrap per box. How many boxes will Mary need to buy to have enough plastic wrap for the week?

    _____

9. An advertising company has 15 different kinds of one-page flyers. The company needs 75 copies of each kind of flyer. How many reams of paper will the company need to produce the flyers? One ream equals 500 sheets of paper.

    _____

TWO-STEP PROBLEMS

Some problems require two steps to solve.

Read each of the following problems carefully and solve.

1. For a family picnic, Renee bought 10 pounds of hamburger meat and used $\frac{1}{4}$ of a pound of meat to make each hamburger patty. Renee's family ate 32 hamburgers. How many pounds of hamburger meat did she have left?

_____

2. Vic sold 45 raffle tickets. His brother sold twice as many. How many tickets did they sell together?

_____

3. Erin earns $2,200 per month. Her deductions amount to 28% of her paycheck. How much does she take home each month?

_____

4. Matheson Middle School Band is selling T-shirts to raise money for new uniforms. They need to raise $1260. They are selling T-shirts for $12 each. There is a $6 profit for each shirt sold. So far, they have sold 85 T-shirts. How many more T-shirts do they need to sell to raise the $1260?

_____

5. Alphonso was earning $1,860 per month and then got a 12% raise. How much will he make per month now?

_____

6. Barbara and Jeff ate out for dinner. The total came to $15.00. They left a 15% tip. How much was the tip and the meal together?

_____

7. Hillary is bicycling across Montana, taking a 845 mile course. The first week, she covered 320 miles. The second week, she traveled another 350 miles. How many more miles does she have to travel to complete the course?

_____

8. Jason budgets 30% of his $1,100 income each month for food. How much money does he have to spend for everything else?

_____

9. After Madison makes a 12% down payment on a $2,000 motorcycle, how much will she still owe?

_____

10. Randy bought a pair of shoes for $51, a tie for $18, and a new belt for $23. If the sales tax is 8%, how much sales tax did he pay?

_____

CHOOSING APPROPRIATE NUMBERS

For each paragraph below, choose the answer from the lettered list that makes the most sense to fill in each blank.

Harrison High School has a total enrollment of (1) _____ students. The students go to class (2) _____ hours per day. Classes end at (3) _____. (4) _____ students ride the bus home.

    A.   3:10
    B   2,698
    C.   1,235
    D.   7

Our dog is (5) _____ years old. Last week, she had a litter of (6) _____ puppies. We should be able to sell her puppies for (7) _____ each. After we pay for puppy shots and food, we should have a profit of (8) _____.

    E. $150
    F.   3
    G. $900
    H.   8

John had a job interview at (9) _____ o'clock this morning. He got the job. Starting pay is (10) _____ dollars an hour. He will work (11) _____ hours per week. His gross earnings should be (12) _____ dollars per week.

    A.   12
    B.   10
    C.   40
    D.   480

Last week, I saw jeans on sale for (13) _____ off the original price of (14) _____. I bought 1 pair and saved (15) _____. If I bought 2 pairs, I would have saved (16) _____.

    E. $40.00
    F. $48.00
    G.   60%
    H. $24.00

On January (17) _____, I will be 16 years old. That means I will be able to get my driver's license. I will need (18) _____ per month for insurance payments for the new car I am buying and (19) _____ for car payments. Gas should cost about (20) _____ per month. Life is going to be expensive!

    A. $50.00
    B. $381.00
    C. $112.00
    D.   20th

It is (21) _____ in the morning. The high today will be (22) _____ above zero. Tonight's low will be (23) _____ below zero. Presently it is a chilly (24) _____ above zero.

    E.   2°
    F.   −5°
    G.   15°
    H.   8:23

USING DIAGRAMS TO SOLVE PROBLEMS

Problems that require logical reasoning cannot always be solved with a set formula. Sometimes, drawing diagrams can help you see the solution.

EXAMPLE: Yvette, Barbara, Patty, and Nicole agreed to meet at the movie theater around 7:00 p.m. Nicole arrived before Yvette. Barbara arrived after Yvette. Patty arrived before Barbara but after Yvette. What is the order of their arrival?

Nicole	Yvette	Patty	Barbara
1st	2nd	3rd	4th

Arrange the names in a diagram so that the order agrees with the problem.

Use a diagram to answer each of the following questions.

1. Javy, Thomas, Pat, and Keith raced their bikes across the playground. Keith beat Thomas but lost to Pat and Javy. Pat beat Javy. Who won the race?

   _____

2. Jeff, Greg, Pedro, Lisa, Macy, and Kay eat lunch together at a round table. Kay wants to sit beside Pedro, Pedro wants to sit next to Lisa, Greg wants to sit next to Macy, and Jeff wants to sit beside Kay. Macy would rather not sit beside Lisa. Which two people should sit on each side of Jeff?

   _____

   _____

3. Three teams play a round-robin tournament where each team plays every other team. Team A beat Team C. Team B beat Team A. Team B beat Team C. Which team is the best?

   _____

4. Caleb, Thomas, Ginger, Alex, and Janice are in the lunch line. Thomas is behind Alex. Caleb is in front of Alex but behind Ginger and Janice. Janice is between Ginger and Caleb. Who is third in line?

   _____

5. Ray, Fleta, Paula, Joan, and Henry hold hands to make a circle. Joan is between Ray and Paula. Fleta is holding Ray's other hand. Paula is also holding Henry's hand. Who must be holding Henry's other hand?

   _____

6. The Bears, the Cavaliers, the Knights, and the Lions all competed in a track meet. One team from each school ran the 400 meter relay race. The Bears beat the Knights but lost to the Cavaliers. The Lions beat the Cavaliers. Who finished first, second, third, and fourth?

   1st _____

   2nd _____

   3rd _____

   4th _____

NUMBER PATTERNS

In each of the examples below, there is a sequence of data that follows a pattern. You must find the pattern that holds true for each number in the data. Once you determine the pattern, you can write out an equation that fits the data and figure out any other number in the sequence.

	1st 5 Numbers in Sequence	Pattern	6th Number in Sequence	20th Number in the Sequence
EXAMPLE 1:	3, 4, 5, 6, 7	$n + 2$	$n + 2 = 8$	$20 + 2 = 22$

In number patterns, the sequence is the output. The input can be the set of whole numbers starting with 1. However, you must determine the "rule" or pattern. Look at the table below.

input	sequence
1 $\longrightarrow$	3
2 $\longrightarrow$	4
3 $\longrightarrow$	5
4 $\longrightarrow$	6
5 $\longrightarrow$	7

What pattern or "rule" can you come up with that gives you the first number in the sequence, 3, when you input 1? $n + 2$ will work because when $n = 1$, the first number in the sequence = 3. Does this pattern hold true for the rest of the numbers in the sequence? Yes, it does. When $n = 2$, the second number in the sequence = 4. When $n = 3$, the third number in the sequence = 5, and so on. Therefore, $n + 2$ is the pattern. Even without knowing the algebraic form of the pattern, you could figure out that 8 is the next number in the sequence. The expression describing this pattern would be $n + 2$. To find the 20th number in the pattern, use $n = 20$ to get 22.

	Sequence	Pattern	Next Number in Sequence	20th Number in the Sequence
EXAMPLE 2:	1, 4, 9, 16, 25	n^2	$n^2 = 36$	400
EXAMPLE 3:	−2, −4, −6, −8, −10	$-2n$	$-2n = -12$	−40

Find the pattern and the next number in each of the sequences below.

	Sequence	Pattern	Next Number in Sequence	20th number in the sequence
1.	−2, −1, 0, 1, 2	_____	_____	_____
2.	5, 6, 7, 8, 9	_____	_____	_____
3.	3, 7, 11, 15, 19	_____	_____	_____
4.	−3, −6, −9, −12, −15	_____	_____	_____
5.	3, 5, 7, 9, 11	_____	_____	_____
6.	2, 4, 8, 16, 32	_____	_____	_____
7.	1, 8, 27, 64, 125	_____	_____	_____
8.	0, −1, −2, −3, −4	_____	_____	_____
9.	2, 5, 10, 17, 26	_____	_____	_____

MAKING PREDICTIONS

Use what you know about number patterns to answer the following questions.

Corn plants grow as tall as they will get in about 20 weeks. Study the chart of the rate of corn plant growth below, and answer the questions that follow.

Corn Growth	
Beginning Week	**Height (inches)**
2	9
7	39
11	63
14	??

1.   If the growth pattern continues, how high will the corn plant be beginning week 14?

2.   If the growth pattern was constant (at the same rate from week to week), how high was the corn in the beginning of the 8th week?

Peter Nichols is staining furniture for a furniture manufacturer. He stains large pieces of furniture that take longer to dry in the beginning of the day and smaller pieces of furniture as the day progresses.

Time	# Pieces Completed per Hour
Hour 1	3
Hour 3	5
Hour 6	8

3.   How many pieces of furniture did Peter stain during his second hour of work?

4.   How many pieces of furniture will Peter have stained by the end of an 8 hour day?

Brian Bailey is bass fishing down the Humbolt River. He has selected six locations to fish. Using his car, he drives to the first location near Golconda. His final location is near Valmy. As he travels south, he notices that the bass catches are getting larger.

Fishing Direction	Fishing Location	Number of bass caught
North	1	4
↓	2	unrecorded
	3	10
↓	4	unrecorded
South	5	16

5.   How many bass would he likely catch in the sixth location?

6.   If he fishes six locations, how many bass is he likely to catch altogether?

INDUCTIVE REASONING AND PATTERNS

Humans have always observed what happened in the past and used these observations to predict what would happen in the future. This is called **inductive reasoning**. Although mathematics is referred to as the "deductive science," it benefits from inductive reasoning. We observe patterns in the mathematical behavior of a phenomenon, and then find a rule or formula for describing and predicting its future mathematical behavior. There are lots of different kinds of predictions that may be of interest.

EXAMPLE 1: Nancy is watching her nephew, Drew, arrange his marbles in rows on the kitchen floor. The figure below shows the progression of his arrangement.

Row 1
Row 2
Row 3
Row 4

QUESTION 1: Assuming this pattern continues, how many marbles would Drew place in a fifth row?

Answer 1: It appears that Drew doubles the number of marbles in each successive row. In the 4th row he had 8 marbles, so in the 5th row we can predict 16 marbles.

QUESTION 2: How many marbles will Drew place in the nth row?

Answer 2: To find a rule for the number of marbles in the nth row, we look at the pattern suggested by the table below.

Which row	1st	2nd	3rd	4th	5th
Number of marbles	1	2	4	8	16

Observing closely, you will notice that the nth row contains 2^{n-1} marbles.

QUESTION 3: Suppose Nancy tells you that Drew now has 6 rows of marbles on the floor. What is the total number of marbles in his arrangement?

Answer 3: Again, organizing the data in a table could be helpful.

Number of rows	1	2	3	4	5
Total number of marbles	1	3	7	15	31

With careful observation, one will notice that the total number of marbles is always 1 less than a power of 2; indeed, for n rows there are $2^n - 1$ marbles total.

QUESTION 4: If Drew has 500 marbles, what is the maximum number of *complete* rows he can form?

Answer 4: With 8 complete rows, Drew will use $2^8 - 1 = 255$ marbles, and to form 9 complete rows he would need $2^9 - 1 = 511$ marbles; thus, the answer is 8 complete rows.

EXAMPLE 2: Manuel drops a golf ball from the roof of his high school while Carla videos the motion of the ball. Later, the video is analyzed, and the results are recorded concerning the height of each bounce of the ball.

QUESTION 1: What height do you predict for the fifth bounce?

Initial height	1st bounce	2nd ounce	3rd bounce	4th bounce
30 ft	18 ft	10.8 ft	6.48 ft	3.888 ft

Answer 1: To answer this question, we need to be able to relate the height of each bounce to the bounce immediately preceding it. Perhaps the best way to do this is with **ratios** as follows:

$$\frac{\text{Height of 1st bounce}}{\text{Initial bounce}} = 0.6 \qquad \frac{\text{Height of 2nd bounce}}{\text{Height of 1st bounce}} = 0.6 \cdots \frac{\text{Height of 4th bounce}}{\text{Height of 3rd bounce}} = 0.6$$

Since the ratio of the height of each bounce to the bounce before it appears constant, we have some basis for making predictions.

Using this, we can reason that the fifth bounce will be equal to 0.6 of the fourth bounce.

Thus, we predict the fifth bounce to have a height of **0.6 × 3.888 = 2.3328 ft.**

QUESTION 2: Which bounce will be the last one with a height of one foot or greater?

Answer 2: For this question, keep looking at predicted bounce heights until a bounce less than 1 foot is reached.

The sixth bounce is predicted to be 1.39968 ft.
The seventh bounce is predicted to be 0.839808 ft.

Thus, the last bounce with a height greater than 1 foot is predicted to be the sixth one.

Read each of the following questions carefully. Use inductive reasoning to answer each question. You may wish to make a table or a diagram to help you visualize the pattern in some of the problems. Show your work.

George is stacking his coins as shown below.

1. How many coins do you predict he will place in the fourth stack?

2. How many coins in an *n*th row?

3. If George has exactly 6 "complete" stacks, how many coins does he have?

4. If George has 2,000 coins, how many complete stacks can he form?

Bob and Alice have designed and created a Web site for their high school. The first week they had 5 visitors to the site; during the second week, they had 10 visitors; and during the third week, they had 20 visitors.

5. If current trends continue, how many visitors can they expect in the fifth week?

6. How many in the *n*th week?

7. How many weeks will it be before they get more than 500 visitors in a single week?

---

8. In 1979 (the first year of classes), there were 500 students at Brookstone High. In 1989, there were 1000 students. In 1999, there were 2000 students. How many students would you predict at Brookstone in 2009 if this pattern continues (and no new schools are built)?

---

9. The number of new drivers' licenses issued in the city of Boomtown, USA was 512 in 1992, 768 in 1994, 1,152 in 1996, and 1,728 in 1998. Estimate the number of new drivers' licenses that will be issued in 2000.

---

10. The average combined (math and verbal) SAT score for seniors at Brookstone High was 1,000 in 1996, 1,100 in 1997, 1,210 in 1998, and 1331 in 1999. Predict the combined SAT score for Brookstone seniors in 2000.

Juan wants to be a medical researcher, inspired in part by the story of how penicillin was discovered as a mold growing on a laboratory dish. One morning, Juan observes a mold on one of his lab dishes. Each morning thereafter, he observes and records the pattern of growth. The mold appeared to cover about 1/32 of the dish surface on the first day, 1/16 on the second day, and 1/8 on the third day.

11. If this rate of growth continues, on which day can Juan expect the entire dish to be covered with mold?

12. Suppose that whenever the original dish gets covered with mold, Juan transfers half of the mold to another dish. How long will it be before *both* dishes are covered again?

13. Every year on the last day of school, the Brookstone High cafeteria serves the principal's favorite dish–Broccoli Surprise. In 1988, 1024 students chose to eat Broccoli Surprise on the last day of school, 512 students in 1992, and 256 students in 1996. Predict how many will choose Broccoli Surprise on the last day of school in 2000.

Part of testing a new drug is determining the rate at which it will break down (*decay*) in the blood. The decay results for a certain antibiotic after a 1000 milligram injection are given in the table below.

12:00 PM	1:00 PM	2:00 PM
1000 mg	800 mg	640 mg

14. Predict the number of milligrams that will be in the patient's bloodstream at 3:00 PM.

15. At which hour can the measurer expect to record a result of less than 300 mg?

16. Marie has a daylily in her mother's garden. Every Saturday morning in the spring, she measures and records its height in the table below. What height do you predict for Marie's daylily on April 29? (Hint: Look at the *change* in height each week when looking for the pattern.)

April 1	April 8	April 15	April 22
12 in	18 in	21 in	22.5 in

17. Bob puts a glass of water in the freezer and records the temperature every 15 minutes. The results are displayed in the table below. If this pattern of cooling continues, what will be the temperature at 2:15 PM? (Hint: Again, look at the *changes* in temperature in order to see the pattern.)

1:00 PM	1:15 PM	1:30 PM	1:45 PM
92°F	60°F	44°F	36°F

Suppose you cut your hand on a rusty nail that deposits 25 bacteria cells into the wound. Suppose also that each bacterium splits into two bacteria every 15 minutes.

18. How many bacteria will there be after two hours?

19. How many 15-minute intervals will pass before there are over a million bacteria?

20. Elias performed a psychology experiment at his school. He found that when someone is asked to pass information along to someone else, only about 70% of the original information is actually passed to the recipient. Suppose Elias gives the information to Brian, Brian passes it along to George, and George passes it to Montel. Using Elias's results from past experiments, what percentage of the original information does Montel actually receive?

MATHEMATICAL REASONING/LOGIC

The ability to use **logic** and **reasoning** is an important skill for solving math problems, but it can also be helpful in real-life situations. For example, if you need to get to Park Street, and the Park Street bus always comes to the bus stop at 3 p.m., then you know that you need to get to the bus stop at least by 3 p.m. This is a real-life example of using logic, which many people would call "common sense."

There are many different types of statements which are commonly used to describe mathematical principles. However, using the rules of logic, the truth of any mathematical statement must be evaluated. Below are a list of tools used in logic to evaluate mathematical statements.

Logic is the discipline that studies valid reasoning. There are many forms of valid arguments, but we will just review a few here.

A **proposition** is usually a declarative sentence which may be true or false.

An **argument** is a set of two or more related propositions, called **premises**, that provide support for another proposition, called the **conclusion**.

Deductive reasoning is an argument which begins with general premises and proceeds to a more specific conclusion. Most elementary mathematical problems use deductive reasoning.

Inductive reasoning is an argument in which the truth of its premises make it likely or probable that its conclusion is true.

ARGUMENTS

Most of logic deals with the evaluation of the validity of arguments. An argument is a group of statements that includes a conclusion and at least one premise. A premise is a statement that you know is true or at least you assume to be true. Then, you draw a conclusion based on what you know or believe is true in the premise(s). Consider the following example:

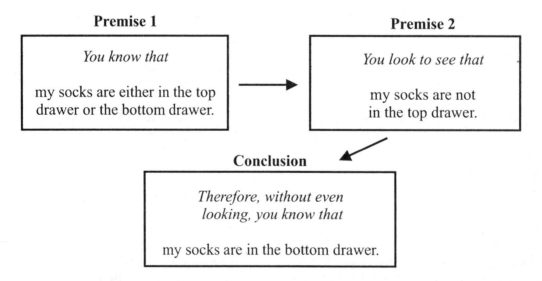

This argument is an example of deductive reasoning, where the conclusion is "deduced" from the premises and nothing else. In other words, if Premise 1 and Premise 2 are true, you don't even need to look in the bottom drawer to know that the conclusion is true.

DEDUCTIVE AND INDUCTIVE ARGUMENTS

In general, there are two types of logical arguments: **deductive** and **inductive**. Deductive arguments tend to move from general statements or theories to more specific conclusions. Inductive arguments tend to move from specific observations to general theories.

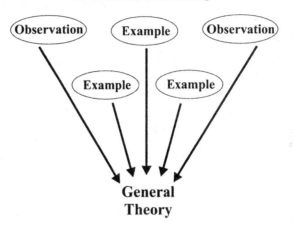

Compare the two examples below:

Deductive Argument

Premise 1	All men are mortal.
Premise 2	Socrates is a man.
Conclusion	Socrates is mortal.

Inductive Argument

Premise 1	The sun rose this morning.
Premise 2	The sun rose yesterday morning.
Premise 3	The sun rose two days ago.
Premise 4	The sun rose three days ago.
Conclusion	The sun will rise tomorrow.

An inductive argument cannot be proved beyond a shadow of a doubt. For example, it's a pretty good bet that the sun will come up tomorrow, but the sun not coming up presents no logical contradiction.

On the other hand, a deductive argument can have logical certainty, but it must be properly constructed. Consider the examples below.

True Conclusion from an Invalid Argument

All men are mortal.
Socrates is mortal.
Therefore, Socrates is a man.

Even though the above conclusion is true, the argument is based on invalid logic. Both men and women are mortal. Therefore, Socrates could be a woman.

False Conclusion from a Valid Argument

All astronauts are men.
Julia Roberts is an astronaut.
Therefore, Julia Roberts is a man.

In this case, the conclusion is false because the premises are false. However, the logic of the argument is valid because *if* the premises were true, then the conclusion would be true.

EXAMPLE 1:   Which argument is valid?

If you speed on Hill Street, you will get a ticket.
If you get a ticket, you will pay a fine.

A.   I paid a fine, so I was speeding on Hill Street.
B.   I got a ticket, so I was speeding on Hill Street.
C.   I exceeded the speed limit on Hill Street, so I paid a fine.
D.   I did not speed on Hill Street, so I did not pay a fine.

Answer: C is valid.
A is incorrect. I could have paid a fine for another violation.
B is incorrect. I could have gotten a ticket for some other violation.
D is incorrect. I could have paid a fine for speeding somewhere else.

EXAMPLE 2:   Assume the given proposition is true. Then determine if each statement is true or false.

Given: If a dog is thirsty, he will drink.
A.   If a dog drinks, then he is thirsty.   T   or   F
B.   If a dog is not thirsty, he will not drink.   T   or   F
C.   If a dog will not drink, he is not thirsty.   T   or   F

Answer: A is false. He is not necessarily thirsty; he could just drink because other dogs are drinking or drink to show others his control of the water. This statement is the converse of the original.
B is false. The reasoning from A applies. This statement is the inverse of the original.
C is true. It is the **contrapositive,** or the complete opposite of the original.

For numbers 1–5, what conclusion can be drawn from each proposition?

1.   All squirrels are rodents. All rodents are mammals. Therefore,

2.   All fractions are rational numbers. All rational numbers are real numbers. Therefore,

3.   All squares are rectangles. All rectangles are parallelograms. All parallelograms are quadrilaterals. Therefore,

4.   All Chevrolets are made by General Motors. All Luminas are Chevrolets. Therefore,

5.   If a number is even and divisible by three, then it is divisible by six. Eighteen is divisible by six. Therefore,

For numbers 6–9, assume the given proposition is true. Then determine if the statements following it are true or false.

All squares are rectangles.
6.   All rectangles are squares.   T   or   F
7.   All non-squares are non-rectangles.   T   or   F
8.   No squares are non-rectangles.   T   or   F
9.   All non-rectangles are non-squares.   T   or   F

CHAPTER 9 REVIEW

Read each problem carefully, and solve using the problem-solving methods you learned in this chapter. If there is not enough information, tell what information is missing.

1. East Point Middle School is taking 321 students, 11 teachers, and 10 parents on a field trip. Each bus holds 45 persons. How many buses will the school need?

   _____

2. Kyle paid for his school lunch with a $10 bill. How much change did he get back?

   _____

3. Tyrone bought shoes for $65 and a shirt for $24. He paid 8% sales tax. What was the total cost of the two items?

   _____

4. Crystal bought a jacket at a 20% off sale. How much did she save buying the jacket on sale?

   _____

5. Laura left her house at 8:30 a.m. in Charlotte, NC and drove 5 hours to see her brother in Atlanta, GA. Her trip from Charlotte to Atlanta totaled 265 miles. What was her average speed in miles per hour?

   _____

6. Nathan drove 4 hours to get home. He arrived at 10:00 p.m. How fast did he drive?

   _____

7. Eight soccer teams make up a conference. Each team plays the others three times during a season. How many games does each team play?

   _____

8. Tran worked 4 days last week. She made $6.26 per hour. Her employer deducted 24% of her pay for taxes. How much was Tran's take-home pay?

   _____

9. Lori bought a sweater originally marked $32.50 and a belt for $12.25. The sweater was on sale for 20% off. How much money did she save by buying the sweater on sale?

   _____

10. Northside High School ordered 556 algebra books from the Galactic Math Company. Thirty-two books fit in a box. How many boxes were needed to ship the order?

    _____

11. Cecilia won 1,390 tickets in the arcade. A stuffed bear costs 434 tickets, a stop watch costs 1,650, and a clown wig costs 1,240 tickets. How many more tickets does Cecilia need to win to purchase the stop watch and the stuffed bear?

    _____

12. Mrs. Rhodes gave each of her 35 second grade students a Valentine's Day card. The cards that she picked out came 12 to a box. How many boxes did she have to purchase?

    _____

13. Seth left his home at 2:00 p.m. He arrived at the airport at 3:30 p.m., $1\frac{1}{2}$ hours before his plane departed. What was Seth's average speed traveling to the airport?

    _____

For each paragraph below, choose the answer from the lettered list that makes the most sense to fill in each blank.

You have until October (14) _____ to turn in your book report. The book report needs to be (15) _____ typewritten pages. The book needs to be at least (16) _____ pages long. The book report will count for (17) _____ of your grade.

    A.   10%
    B.   2
    C.   250
    D.   30

I went to the movie with (18) _____ friends last night. The movie started at (19) _____. We each paid (20) _____ for the movie. We got back about (21) _____.

    E.   10:30
    F.   7:00
    G. $7.50
    H.   3

At the beginning of this trip, Ryan's odometer read (22) _____. At the end of his trip, it read (23) _____. The trip took (24) _____ hours. He arrived at (25) _____.

    A.   9:00
    B.   35,758.9
    C.   36,025.9
    D.   6

26. David found a motorcycle he wants to buy that is priced at $4,800. If he waits to buy it at the end of the year, it will be on sale for 20% off. How much will it cost then?

_____

27. Movie tickets sell for $7.50 each. On Wednesday nights tickets are 30% off the regular price. How much would two tickets costs on Wednesday night?

_____

28. A living room set can be purchased for $1,200 or 100 down and $120 a month for 12 months. How much less is the cash price than the installment plan?

_____

29. Janelle drove an average of 50 miles per hour for 275 miles. How many hours did the trip last?

_____

30. Floor tiles are sold only in boxes of 11 tiles per box. Christy needs 140 tiles for her kitchen floor. How many boxes of tile will she need to buy?

_____

31. Kyle works 40 hours per week for $11.20 per hour. His total deductions come to $37.80. What is his take-home pay?

_____

32. Vince, Hal, Weng, and Carl raced on roller blades down a hill. Vince beat Carl. Hal finished before Vince but after Weng. Who won the race?

33. Mr. Sanders is hiring someone for the newest position that has opened up at his company. He interviewed four people for the job: Rick, Luca, Janelle, and Jacob. To help with his decision, he puts the four candidates in order from best qualified to least qualified. Janelle and Rick are better qualified than Luca. Rick and Luca are better qualified than Jacob, but Luca is less qualified than Janelle. If Janelle is not as qualified for the position as Rick, who is Mr. Sanders going to hire for the new opening in his company?

Find the pattern for the following number sequences, and then find the *n*th number requested.

34. 0, 1, 2, 3, 4
pattern _____
20th number _____

35. 1, 3, 5, 7, 9
pattern _____
25th number _____

36. 3, 6, 9, 12, 15
pattern _____
30th number _____

Olivia starts, maintains, and sells ant farms as a hobby. She had 500 ants in 1997, 2,000 in 1998, and 8,000 in 1999.

37. If her hobby continued to grow as it had since 1997, how many ants did Olivia have in 2003?

38. How many in the *n*th year after 1997?

39. In what year would she have more than 100,000 ants?

Justin has just gotten a bill from his Internet Service Provider. The first four months of charges for his service are recorded in the table below.

	January	February	March	April
Hours	0	10	5	25
Charge	$4.95	$14.45	$9.70	$28.70

40. Write a formula for the cost of *n* hours of Internet service.

41. What is the greatest number of hours he can be on the Internet and still keep his bill under $20.00?

Lisa is baking cookies for the Fall Festival. She bakes 27 cookies with each batch of batter. However, she has a defective oven, which results in 5 cookies in each batch being burned.

42. Write a formula for the number of cookies available for the festival as a result of Lisa baking *n* batches of cookies.

43. How many batches does she need in order to produce 300 cookies for the festival?

44. How many cookies (counting burned ones) will she actually bake?

Chapter 10

MEAN AND MEDIAN

MEAN

In statistics, the **mean** is the same as the **average**. To find the **mean** of a list of numbers, first add together all the numbers in the list, and then divide by the number of items in the list.

EXAMPLE: Find the mean of : 38, 72, 110, 548

Step 1: First Add: $38 + 72 + 110 + 548 = \mathbf{768}$

Step 2: There are 4 numbers in the list, so divide the total by 4. $4\overline{)768}$ $\begin{array}{r} 192 \\ \hline \end{array}$
The mean is 192.

Practice finding the mean (average). Round to the nearest tenth if necessary.

1. Dinners served:

    489 561 522 450

    Mean = _____

2. Prices paid for shirts:

    $4.89 $9.97 $5.90 $8.64

    Average = _____

3. Piglets born

    23 19 15 21 22

    Mean = _____

4. Student Absences:

    6 5 13 8 9 12 7

    Average = _____

5. Paychecks:

    $89.56 $99.99 $56.54

    Average = _____

6. Choir attendance:

    56 45 97 66 70

    Mean = _____

7. Long distance calls:

    33 14 24 21 19

    Mean = _____

8. Train boxcars:

    56 55 48 61 51

    Mean = _____

9. Cookies eaten:

    5 6 8 9 2 4 3

    Mean = _____

Find the mean (average) of the following word problems.

10. Val's science grades were: 95, 87, 65, 94, 78, and 97. What was her average? _____

11. Ann runs a business from her home. The number of orders for the last 7 business days were: 17, 24, 13, 8, 11, 15, and 9. What was the average number of orders per day? _____

12. Melissa tracked the number of phone calls she had per day: 8, 2, 5, 4, 7, 3, 6, 1. What was the average number of calls she received? _____

MEDIAN

In a list of numbers ordered from lowest to highest, the **median** is the middle number. To find the **median,** first arrange the numbers in numerical order. If there is an odd number of items in the list, the **median** is the middle number. If there is an even number of items in the list, the **median** is the **average of the two middle numbers.**

EXAMPLE 1: Find the median of: 42, 35, 45, 37, and 41.

Step 1: Arrange the numbers in numerical order: 35 37 ㊶ 42 45

Step 2: Find the middle number. **The median is 41.**

EXAMPLE 2: Find the median of 14, 53, 42, 6, 14, and 46.

Step 1: Arrange the numbers in numerical order: 6 14 ⟮14 42⟯ 46 53

Step 2: Find the average of the 2 middle numbers.
$(14 + 42) \div 2 = 28.$ **The median is 28.**

Circle the median in each list of numbers.

1. 35, 55, 40, 30, and 45   4. 15, 16, 19, 25, and 20   7. 401, 758, and 254

2. 7, 2, 3, 6, 5, 1, and 8   5. 75, 98, 87, 65, 82, 88, and 100   8. 41, 23, 14, 21, and 19

3. 65, 42, 60, 46, and 90   6. 33, 42, 50, 22, and 19   9. 5, 8, 3, 10, 13, 1, and 8

10.	11.	12.	13.	14.	15.	16.
19	9	45	52	20	8	15
14	3	32	54	21	17	40
12	10	66	19	25	13	42
15	17	55	63	18	14	32
18	6	61	20	16	22	28

Find the median in each list of numbers.

17. 10, 8, 21, 14, 9, and 12 _____   20. 48, 13, 54, 82, 90, and 7 _____

18. 43, 36, 20, and 40 _____   21. 23, 21, 36, and 27 _____

19. 5, 24, 9, 18, 12, and 3 _____   22. 9, 4, 3, 1, 6, 2, 10, and 12 _____

23.	24.	25.	26.	27.	28.	29.
2	11	13	75	48	22	17
10	22	15	62	45	19	30
6	25	9	60	52	15	31
18	28	35	52	30	43	18
20	10	29	80	35	34	14
23	23	33	50	58	28	25

_____   _____   _____   _____   _____   _____   _____

MISLEADING STATISTICS

As you read magazines and newspapers, you will see many charts and graphs which present statistical data. This data will illustrate how measurements change over time or how one measurement corresponds to another measurement. However, some charts and graphs are presented to make changes in data appear greater than they actually are. The people presenting the data create these distortions to make exaggerated claims.

There is one method to arrange the data in ways which can exaggerate statistical measurements. A statistician can create a graph in which the number line does not begin with zero.

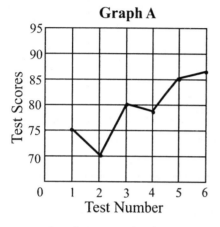

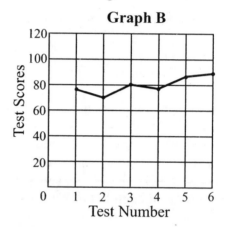

In the two graphs above, notice how each graph displays the same data. However, the way the data is displayed in graph A appears more striking than the data display in graph B. Graph A's data presentation is more striking because the test score numbers do not begin at zero.

Another form of misleading information is through the use of the wrong statistical measure to determine what is the middle. For instance, the mean, or average, of many data measurements allows **outliers** (data measurements which lie well outside the normal range) to have a large effect. Examine the measurements in the chart below.

Address	Household Income	Address	Household Income
341 Spring Drive	$19,000	346 Spring Drive	$30,000
342 Spring Drive	$17,000	347 Spring Drive	$32,000
343 Spring Drive	$26,000	348 Spring Drive	$1,870,000
344 Spring Drive	$22,000	349 Spring Drive	$31,000
345 Spring Drive	$25,000	350 Spring Drive	$28,000

Average (Mean) Household Income: $210,000
Median Household Income: $27,000

In this example, the outlier, located at 348 Spring Drive, inflates the average household income on this street to the extent that it is over eight times the median income for the area.

Read the following charts and graphs, and then answer the questions below.

Graph A - Stasia's Weight Loss

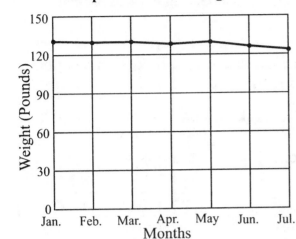

Graph B - Stasia's Weight Loss

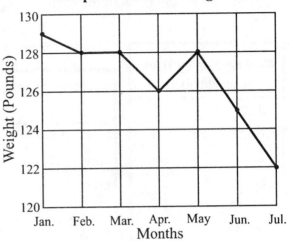

1. Which graph above presents misleading statistical information? Why is the graph misleading?

Twenty teenagers were asked how many electronic and computer games they purchased per year. The following table shows the results.

Number of Games	0	1	2	3	4	5	58
Number of Teenagers	4	2	5	3	4	1	1

2. Find the mean of the data.
3. Find the median of the data.
4. Which measurement is most misleading?
5. Which measurement would depict the data most accurately?
6. Is the *mean* of a set of data affected by outliers? Justify your answer with the example above.

Examine the two bar graphs below.

A. **Avg. Daily Temp.(°F)**

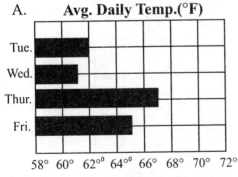

B. **Avg. Daily Temp.(°F)**

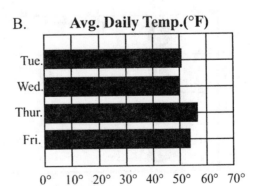

7. Which graph is misleading? Why?

Find the mean and median for each of the following sets of data. Fill in the table below.

❶ Miles Run by Track Team Members

Jeff	24
Eric	20
Craig	19
Simon	20
Elijah	25
Rich	19
Marcus	20

❷ 1992 SUMMER OLYMPIC GAMES
Gold Medals Won

Unified Team	45	Hungary	11
United States	37	South Korea	12
Germany	33	France	8
China	16	Australia	7
Cuba	14	Japan	3
Spain	13		

❸ Hardware Store Payroll June Week 2

Erica	$280
Dane	$206
Sam	$240
Nancy	$404
Elsie	$210
Gail	$305
David	$280

❹ Lunches Served

Fairfield's	80
Mikey's	106
House of China	54
Anticoli's	105
Rib Ranch	83
Two Brothers	66
Mountain Inn	80

❺ Average Days of Rain or Snow BIRMINGHAM

January	12	July	12
February	10	August	11
March	11	September	8
April	11	October	6
May	9	November	8
June	11	December	11

❻ PHONE BILLS

January	$80.50
February	$71.97
March	$82.02
April	$80.50
May	$98.19
June	$108.82

Data Set Number	Mean	Median
❶		
❷		
❸		
❹		
❺		
❻		

7. How do outliers affect the mean in a set of data? Why is this a bad thing?

Chapter 11

DATA INTERPRETATION

READING TABLES

A **table** is a concise way to organize large quantities of information using rows and columns. **Read each table carefully, and then answer the questions that follow.**

Some employers use a tax table like the one below to figure how much **Federal Income Tax** should be withheld from a single person paid weekly. The number of withholding allowances claimed is also commonly referred to as the number of deductions claimed.

Federal Income Tax Withholding Table SINGLE Persons – WEEKLY Payroll Period					
If the wages are –		And the number of withholding allowances claimed is –			
		0	1	2	3
At least	But less than	The amount of income tax to be withheld is –			
$250	260	31	23	16	9
$260	270	32	25	17	10
$270	280	34	26	19	12
$280	290	35	28	20	13
$290	300	37	29	22	15

1. David is single, claims 2 withholding allowances, and earned $275 last week. How much Federal Income Tax was withheld? _____

2. Cecily earned $297 last week and claims 0 deductions. How much Federal Income Tax was withheld? _____

3. Sherri claims 3 deductions and earned $268 last week. How much Federal Income Tax was withheld from her check? _____

4. Mitch is single and claims 1 allowance. Last week, he earned $291. How much was withheld from his check for Federal Income Tax? _____

5. Ginger earned $275 this week and claims 0 deductions. How much Federal Income Tax will be withheld from her check? _____

6. Bill is single and earns $263 per week. He claims 1 withholding allowance. How much Federal Income Tax is withheld each week? _____

Nutritional tables appear on the packaging of nearly every food we eat. Use the nutritional table below to answer questions 1–6.

Peanut Butter Spread

Nutrition Facts	Amount/serving		%DV*	
Serv. Size 2 tbsp (36 g)	**Total Fat**	12 g	**18%**	Vitamin A 0%
Servings Per Container about 14	Sat. Fat	2.5 g	**12%**	Vitamin C 0%
				Calcium 2%
Calories 190	**Cholest.**	0 mg	**0%**	Iron 4%
Calories from Fat 100	**Sodium**	210 mg	**9%**	Niacin 25%
* Percent Daily Values	**Total Carb.**	5 g	**5%**	Vitamin B$_6$ 6%
(DV) are based on a	Fiber	1 g	**5%**	Folate 6%
2,000 calorie diet	Sugars	5 g		Magnesium 15%
	Protein	9 g		Zinc 6%
				Copper 10%

1. How many calories are in 1 tablespoon of peanut butter spread?   _____

2. How many milligrams of sodium are in 4 tablespoons of peanut butter spread?   _____

3. How many grams of saturated fat are in 6 tablespoons of peanut butter spread?   _____

4. If you eat 1 tablespoon of peanut butter spread, what percent of daily value of copper would you eat?   _____

5. If you ate 4 tablespoons of peanut butter spread, how many grams of protein would you get?   _____

6. If you used 2 tablespoons of peanut butter spread to make a peanut butter and jelly sandwich, how many grams of sugar came from the peanut butter spread?   _____

Use the following life expectancy table to answer questions 7–12.

Average Life Expectancy		
Country	Male	Female
Bolivia	45	51
United Kingdom	72	77
India	46	43
Japan	74	81
Peru	52	55
Sweden	73	80
USA	71	78

7. Where do males live longer than females?   _____

8. Where do women live the longest?   _____

9. How much longer do men in the USA live than the men in Peru?   _____

10. Where do men have the shortest life?   _____

11. How much longer does a female in Sweden live than a male in Bolivia?   _____

12. How much shorter is the average life of a female in the USA than a female in Japan?   _____

160

MILEAGE CHART

A mileage chart will tell you the distance between two cities. To read the chart, find the city you start from along the side of the chart. Find the city of your destination along the top. Then read down the column and across the row to see where they intersect. The box where they intersect will tell how many miles it is between the two cities.

	Atlanta	Boise	Columbia	Dallas	Denver	Jacksonville	Kansas City	Little Rock	Miami	New Orleans	Reno	Seattle
Atlanta	0	1750	200	750	1125	275	650	462	550	400	1875	2075
Boise	1750	0	1925	1200	625	2000	1075	1300	2200	1675	300	400
Columbia	200	1925	0	975	1300	250	800	650	550	600	2050	2200
Dallas	750	1200	975	0	625	950	500	350	1100	500	1250	1600
Denver	1125	625	1300	625	0	1400	500	700	1600	1025	750	1000
Jacksonville	275	2000	250	950	1400	0	950	700	300	500	2125	2350
Kansas City	650	1075	800	500	500	950	0	300	1200	700	1225	1400
Little Rock	462	1300	650	350	700	700	300	0	900	400	1425	1675
Miami	550	2200	550	1100	1600	300	1200	900	0	600	2350	2600
New Orleans	400	1675	600	500	1025	500	700	400	600	0	1725	2025
Reno	1875	300	2050	1250	750	2125	1225	1425	2350	1725	0	550
Seattle	2075	400	2200	1600	1000	2350	1400	1675	2600	2025	550	0

Find the distance between the following cities.

1. Little Rock to Dallas _____

2. Seattle to Jacksonville _____

3. Miami to Seattle _____

4. Kansas City to New Orleans _____

5. Denver to Dallas _____

6. Reno to Boise _____

7. Atlanta to Reno _____

8. Little Rock to Columbia _____

9. Seattle to Kansas City _____

10. Columbia to Reno _____

11. Jacksonville to Little Rock _____

12. Columbia to Denver _____

13. New Orleans to Miami _____

14. Reno to Jacksonville _____

15. Miami to Boise _____

16. Jacksonville to Denver _____

17. Kansas City to Columbia _____

18. Dallas to Seattle _____

19. Denver to Miami _____

20. Atlanta to Seattle _____

BAR GRAPHS

Bar graphs can be either vertical or horizontal. There may be just one bar or more than one bar for each interval. Sometimes each bar is divided into two or more parts. In this section, you will work with a variety of bar graphs. Be sure to read all titles, keys, and labels to completely understand all the data that is presented. **Answer the questions about each graph below.**

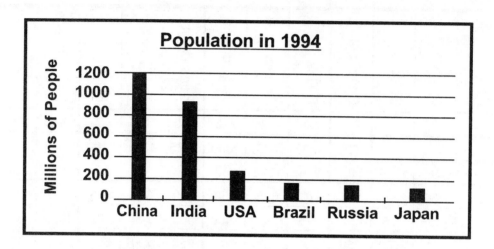

1. Which country has over 1 billion people? _____

2. How many countries have fewer than 200,000,000 people? _____

3. How many more people does India have than Japan? _____

4. If you added together the populations of the USA, Brazil, Russia, and Japan, would it come closer to the population of India or China? _____

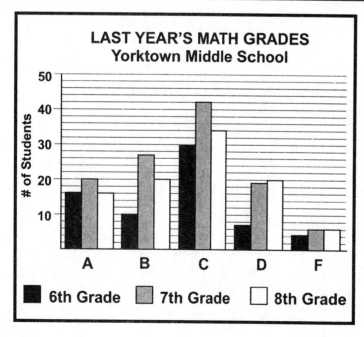

5. How many of last year's 6th graders made C's in math? _____

6. How many more math students made B's in the 7th grade than in the 8th grade? _____

7. How many students at Yorktown Middle school made D's in math last year? _____

8. How many 8th graders took math last year? _____

9. How many students made A's in math last year? _____

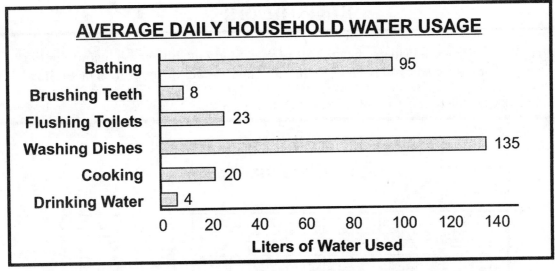

AVERAGE DAILY HOUSEHOLD WATER USAGE

Bathing — 95
Brushing Teeth — 8
Flushing Toilets — 23
Washing Dishes — 135
Cooking — 20
Drinking Water — 4

Liters of Water Used

1. How many liters of water does the average household use every day by washing dishes and cooking? _____

2. How many liters of water would an average household drink in a week (7 days)? _____

3. How many liters of water does an average household use in a day? _____

4. How many more liters does an average household use for bathing than it does for cooking? _____

5. How many liters does the average household use every day to bathe and brush teeth? _____

An elementary school planted four trees: a dogwood, a maple, a pine, and an oak. All four trees were 36 inches tall when planted. The students monitored the growth of the trees each season for one year. Study the graph below, and then answer the questions.

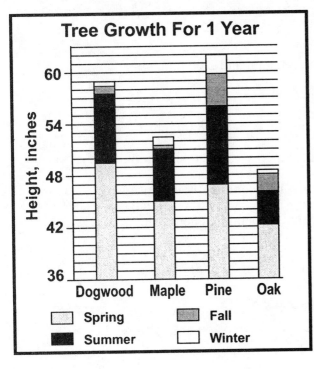

Tree Growth For 1 Year

6. How many more inches did the pine tree grow than the oak tree in the summer? _____

7. Which tree grew the most during the winter? _____

8. Which tree grew the least during the fall? _____

9. How tall was the dogwood after one year? _____

10. How many more inches did the maple tree grow than the oak tree during the whole year? _____

LINE GRAPHS

Line graphs often show how data changes over time. Study the line graph below charting temperature changes for a day in Atlanta, Georgia. Then answer the questions that follow.

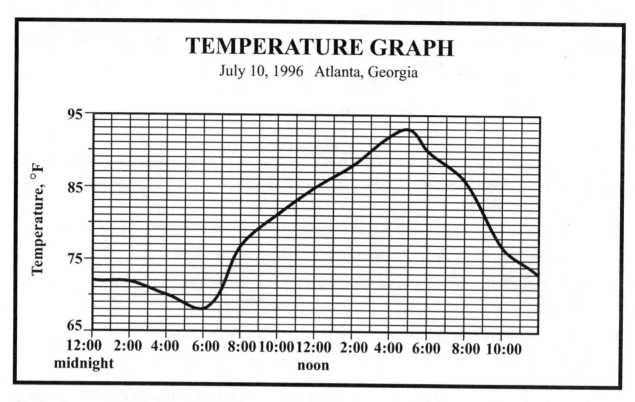

Study the graph and then answer the questions below.

1. When was the coolest time of the day?

2. When was the hottest time of the day?

3. How much did the temperature rise between 6:00 a.m. and 2:00 p.m.?

4. How much did the temperature drop between 6:00 p.m. and 11:00 p.m.?

5. What is the difference in temperature between 8:00 a.m. and 8:00 p.m.?

6. In which two hour period was there the greatest increase in temperature?

7. Between which hours of the day did the temperature continually increase?

8. In which two hour period during the day did the temperature change the least?

9. How much did the temperature decrease from 2:00 a.m. to 6:00 a.m.?

10. During which two times of day was the temperature 88°F?

MULTIPLE LINE GRAPHS

Multiple line graphs are a way to present a large quantity of data in a small space. It would often take several paragraphs to explain in words the same information that one graph could do.

On the graph below, there are three lines. You will need to read the **key** to understand the meaning of each.

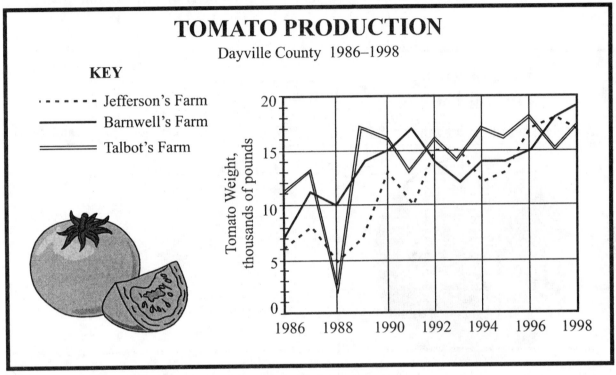

TOMATO PRODUCTION

Dayville County 1986–1998

KEY

- - - - - - Jefferson's Farm
———— Barnwell's Farm
═════ Talbot's Farm

Tomato Weight, thousands of pounds

Study the graph, and then answer the questions below.

1. In what year did Barnwell's Farm produce 8,000 pounds of tomatoes more than Talbot's Farm? _____

2. In which year did Dayville County produce the most pounds of tomatoes? _____

3. In 1991, how many more pounds of tomatoes did Barnwell's Farm produce than Talbot's Farm? _____

4. How many pounds of tomatoes did Dayville County's three farms produce in 1990? _____

5. In which year did Dayville County produce the fewest pounds of tomatoes? _____

6. Which farm had the most dramatic increase in production from one year to the next? _____

7. How many more pounds of tomatoes did Jefferson's Farm produce in 1990 than in 1986? _____

8. Which farm produced the most pounds of tomatoes in 1993? _____

CIRCLE GRAPHS

Circle graphs represent data expressed in percentages of a total. The parts in a circle graph should always add up to 100%. Circle graphs are sometimes called **pie graphs** or **pie charts**.

To figure the value of a percent in a circle graph, multiply the percent by the total. Use the circle graphs below to answer the questions. The first question is worked for you as an example.

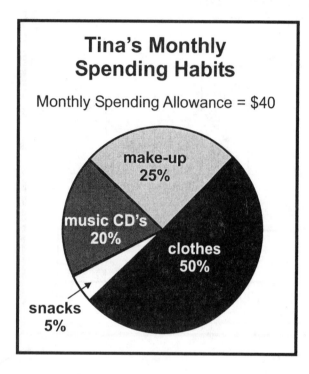

1. How much did Tina spend each month on music CD's?

   $40 × 0.20 = $8.00

   $8.00

2. How much did Tina spend each month on make-up?

   _____

3. How much did Tina spend each month on clothes?

   _____

4. How much did Tina spend each month on snacks?

   _____

Fill in the following chart.

Favorite Activity	Number of students
5. watching TV	1000 × 0.3 = 300
6. talking on the phone	
7. playing video games	
8. surfing the Internet	
9. playing sports	
10. reading	

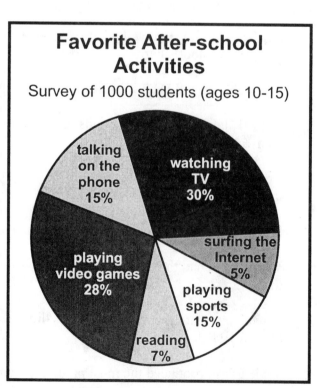

PICTOGRAPHS

Pictographs represent data using symbols. The **key** or **legend** tells what the symbol stands for. Before answering questions about the graph, be sure to read the title, key, and horizontal and vertical labels.

Number of Military Officers by Branch of Service

U.S. Bureau of the Census, 1994

Key: Each military symbol = 10,000 officers

Answer each question below.

1. How many military officers are in the Marines? _____

2. Which branch of the service has the least number of officers? _____

3. Are there more Air Force officers or Army officers? _____

4. How many military officers are there in all? _____

5. How many officers do the Navy and Marines have altogether? _____

6. How many more Navy officers are there than Marine officers? _____

7. How many Army and Coast Guard officers are there altogether? _____

WORLD PRODUCTION AND CONSUMPTION OF PETROLEUM

● Production   ○ Consumption

Each symbol signifies one million barrels per day

Middle East
●●●●●●●●●●●●●●●●●●●●●(
○○○○◐

Europe
●●●●●●●●●●●●●●
○○○○○○○○○○○○○○○○○○○○○○

United States and Canada
●●●●●●●●●(
○○○○○○○○○○○○○○○○○○

Latin America
●●●●●●●●●
○○○○○◐

Asia
●●●●●●(
○○○○○○○○○○○○◐

Africa
●●●●●●
○◐

Australia & Pacific islands
(
◑

1. What part of the world consumes the most oil? _____

2. How many parts of the world produce more oil than they consume? _____

3. How many parts of the world consume more oil than they produce? _____

4. Which part of the world produces three times more oil than it consumes? _____

5. How many more barrels of oil does Europe produce daily than Africa? _____

6. How many barrels does the Middle East produce every day? _____

7. Which areas of the world consume more than twice as much oil as they produce? _____

CATALOG ORDERING

Follow directions for catalog ordering carefully.

DOOGIE'S DOG TREATS
The treats dogs beg for!

Doogie's dog treats are made with only the finest ingredients. They contain real meat and are low in fat with no added sugar or salt.

Flavors:

Cheese	**Beef**
Lamb & Rice	**Peanut Butter**

2 lb bag of any 1 flavor ... $4.75
4 lb bag of any 1 flavor ... $6.50
20 lb bag of any 1 flavor ... $25.25

Shipping & Handling
If total order is:

$0 to $10.00 add $3.75
$10.01 to $20.00 add ... $4.25
$20.01 to $30.00 add ... $5.75
$30.01 & Over add $6.25

Figure the total amount to send with each of the following Doogie Dog Treat orders.

1. one 2-lb bag of beef flavored treats

    _____

2. one 20-lb bag of beef treats, two 4 lb bags of cheese treats, and one 2 lb bag of peanut butter treats

    _____

3. two 2-lb bags of lamb & rice treats, two 4 lb bags of beef treats, one 4 lb bag of cheese treats

    _____

4. one 20-lb bag of each flavor

    _____

5. one 4-lb bag of each flavor

    _____

Figure the total amount to send with each of the following CD Bargain Cellar orders. Don't forget to take discounts when possible.

6. 2 new release rock CD's and 4 "classic" rock CD's   _____

7. 5 "classic" country CD's   _____

8. 1 new release pop CD, 1 new release rap CD, 1 "classic" gospel CD   _____

9. 4 "classic" country CD's and 1 new release country CD   _____

PANCAKE CAFÉ

CLASSIC BREAKFASTS

Bacon or Sausage Add $1.90   Ham Add $1.80   Hash Browns Add $1.20
Orange Juice Add $2.00   Coffee Add $1.20   Tea Add $1.20

Pancakes		Omelets		Sampler Breakfasts	
Buttermilk (4)	$3.40	Country Omelet	$6.00	2 Eggs, 2 Pancakes, 2 Bacon	$5.50
(2)	$2.50	Vegetable Omelet	$5.50	2 Eggs, Hash Browns, 2 Sausage	$5.50
Pecan	$4.00	Steak Omelet	$6.50	2 Eggs, 2 Biscuits, 2 Bacon	$5.50
Chocolate Chip	$4.00	Denver Omelet	$5.70	Eggs Benedict, Toast, Hash Browns	$5.70
Blueberry	$3.50	Sausage Omelet	$6.60	Steak and Eggs, Toast, Hash Browns	$6.50
Buckwheat	$4.00	Cheese Omelet	$6.20	French Toast, 2 Bacon, 2 Sausage	$5.80

Figure the total cost of the following orders.   Round to the nearest penny.

Total Cost

1. Vegetable omelet, hash browns, orange juice, coffee + 5% sales tax _____

2. Chocolate chip pancakes, orange juice, ham + 6% sales tax _____

3. Steak omelet and tea + 7% sales tax _____

4. 2 eggs, 2 biscuits, 2 bacon, coffee + 5% sales tax _____

5. French toast, 2 bacon, 2 sausage, coffee + 4% sales tax _____

6. Denver omelet, hash browns, orange juice + 5% sales tax _____

7. Buckwheat pancakes, ham, coffee, orange juice + 6% sales tax _____

8. 2 eggs, 2 pancakes, 2 bacon, orange juice, coffee + 4% sales tax _____

9. 2 buttermilk pancakes, sausage, coffee + 7% sales tax _____

10. Eggs Benedict, toast, hash browns, tea + 6% sales tax _____

11. Cheese omelet, orange juice, bacon, coffee + 5% sales tax _____

12. Blueberry pancakes, sausage, orange juice, coffee _____

CHAPTER 11 REVIEW

KNIGHTS BASKETBALL Points Scored				
Player	game 1	game 2	game 3	game 4
Joey	5	2	4	8
Lennard	10	8	10	12
Myron	2	6	5	6
Ned	1	3	6	2
Phil	0	4	7	8
Warren	7	2	9	4
Zeek	8	6	7	4

1. How many points did the Knights basketball team score in game 1?

_____

2. How many more points did Warren score in game 3 than in game 1?

_____

3. How many points did Lennard score in the first 4 games?

_____

Use the following catalog information to answer questions 4 & 5.

4. Tommy wants to order 1 basic model airplane set and 2 deluxe sets. How much money should he send with his order?

_____

5. How much would it cost to order 2 standard model airplane sets and 1 deluxe set?

_____

Model Airplanes
B1 bombers, F14's, F15's, F16's, and more!

Basic $24.90
(includes only building materials)

Standard $29.90
(includes building materials plus paint)

Deluxe $35.90
(includes building materials, paint, and decals plus a pamphlet on the history of the aircraft)

Add 10% for shipping and handling

HAPPY'S ICE-CREAM SHOP

Scoops

Cup, cake cone, or sugar cone:
Single $1.10
Double ... $1.75
Triple $2.25

Waffle Cone: Add 50¢

Each topping / Additional toppings: Add 50¢

Sundaes

includes: Ice-cream, 2 toppings, Whipped cream, and Cherry
Child $1.50
Small $2.10
Regular $2.90
Large $3.55

Flavors: Chocolate, Vanilla, Strawberry, Chocolate Chip, Tropical, Black Cherry, Peach, Butter Pecan, Banana Nut

Toppings: Hot fudge, Hot caramel, Hot butterscotch, Almonds Chocolate chips, Sprinkles, Wet walnuts, Peanut butter pieces

6. Sheldon ordered a regular sundae with chocolate ice-cream, hot fudge, and almonds. He chose an additional topping of sprinkles. What was the cost of his sundae?

_____

7. Klarissa ordered a double scoop of chocolate chip ice-cream in a waffle cone with peanut butter pieces. How much was her order?

_____

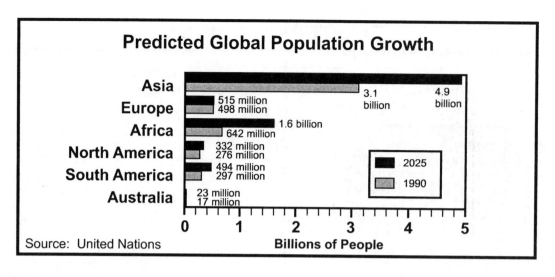

Predicted Global Population Growth

Asia — 515 million / 498 million / 3.1 billion / 4.9 billion
Europe — 515 million / 498 million
Africa — 1.6 billion / 642 million
North America — 332 million / 276 million
South America — 494 million / 297 million
Australia — 23 million / 17 million

Legend: 2025 / 1990

Source: United Nations

Billions of People (0 1 2 3 4 5)

8. By how many is Asia's population predicted to increase between 1990 and 2025?

_____

9. In 1990, how much larger was Africa's population than Europe's?

_____

10. Between 1990 and 2025, how much is Africa's population predicted to increase?

_____

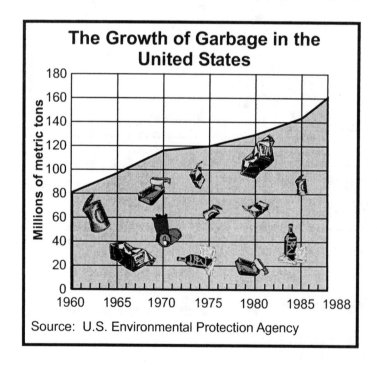

The Growth of Garbage in the United States

Millions of metric tons (0, 20, 40, 60, 80, 100, 120, 140, 160, 180)

1960 1965 1970 1975 1980 1985 1988

Source: U.S. Environmental Protection Agency

11. How much did the volume of garbage grow between 1960 and 1988?

_____

12. In which year did garbage in the United States reach 140 million metric tons?

_____

13. How much did the volume of garbage grow between 1960 and 1966?

_____

172

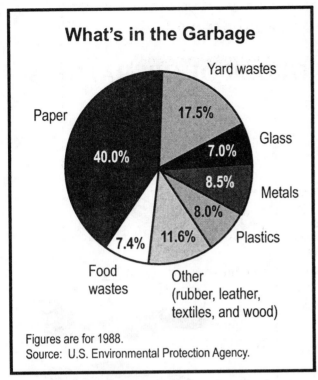

What's in the Garbage

Paper 40.0%
Yard wastes 17.5%
Glass 7.0%
Metals 8.5%
Plastics 8.0%
Other (rubber, leather, textiles, and wood) 11.6%
Food wastes 7.4%

Figures are for 1988.
Source: U.S. Environmental Protection Agency.

14. In 1988, the United States produced 160 million metric tons of garbage. According to the pie chart, how much glass was in the garbage?

_____

15. Out of the 160 million metric tons of garbage, how much was glass, plastic, and metal?

_____

16. If in 1990, the garbage reached 200 million metric tons, and the percentage of wastes remained the same as in 1988, how much food would have been in the 1990 garbage?

_____

The students at a middle school raised money for a school-wide field day by selling candy bars for a month. The graph below shows sales per grade each week.

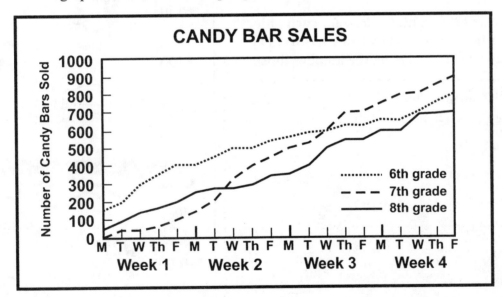

CANDY BAR SALES

Number of Candy Bars Sold

.... 6th grade
- - - - 7th grade
——— 8th grade

Week 1   Week 2   Week 3   Week 4

17. On Friday of week 1, how many more candy bars had the 6th grade sold compared to the 7th grade?

_____

18. By the end of week 3, which grade had sold the most candy bars?

_____

19. On Thursday of week 2, how many candy bars had the 6th, 7th, and 8th grades sold?

_____

20. How many total candy bars did all three grades sell by the end of week 4? _____

173

INTEGERS AND ORDER OF OPERATIONS

In elementary school, you learned to use whole numbers.

Whole numbers = { 0, 1, 2, 3, 4, 5 . . . }

For most things in life, whole numbers are all we need to use. However, when your checking account falls below zero or the temperature falls below zero, we need a way to express that. Mathematicians have decided that a negative sign, which looks exactly like a subtraction sign, would be used in front of a number to show that the number is below zero. All the negative whole numbers and positive whole numbers plus zero make up the set of integers.

Integers = { . . . –4, –3, –2, –1, 0, 1, 2, 3, 4 . . . }

For pictures at the right, tell how far each object is above or below sea level by using positive (+) and negative (–) numbers on the number line.

1. Nuclear Submarine _____

2. F-14A (black) _____

3. school of fish _____

4. Titan Rocket _____

5. whale _____

6. F-15 (gray) _____

7. seagulls _____

8. shells _____

9. Chinook Helicopter _____

10. What if the whale went up 50 feet (+50)? Where would it be? _____

11. What if the submarine went down 50 feet (–50)? Where would it be? _____

12. What if the seahorse went up 20 feet? Where would it be then? _____

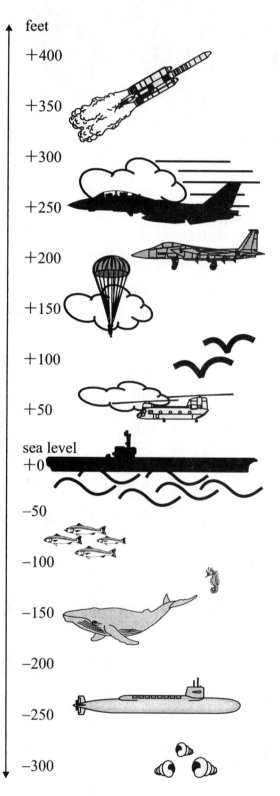

ABSOLUTE VALUE

The absolute value of a number is the distance the number is from zero on the number line.

The absolute value of 6 is written | 6 |. | 6 | = 6
The absolute value of −6 is written | −6 |. | −6 | = 6

Both 6 and −6 are the same distance, 6 spaces, from zero so their absolute value is the same, 6.

EXAMPLES:

| −4 | = 4   − | −4 | = −4 | −9 | + 5 = 9 + 5 = 14

| 9 | − | 8 | = 9 − 8 = 1 | 6 | − | −6 | = 6 − 6 = 0 | −5 | + | −2 | = 5 + 2 = 7

Simplify the following absolute value problems.

1. | 9 | = _____   6. | −4 | − | 3 | = _____   11. | 7 | − | −5 | = _____

2.   − | 5 | = _____   7. | −8 | − | −4 | = _____   12. | −12 | + | −2 | = _____

3. | −25 | = _____   8. | 5 | + | −4 | = _____   13. | 15 | − | 4 | = _____

4.   − | −12 | = _____   9. | −2 | + | 6 | = _____   14. | −6 | + | 8 | = _____

5.   − | 64 | = _____   10. | 10 | + | 8 | = _____   15. | 17 | − | −5 | = _____

ADDING INTEGERS

First, we will see how to add integers on the number line; then we will learn rules for working the problems without using a number line.

EXAMPLE 1: Add: (−3) + 7

Step 1:   The first integer in the problem tells us where to start.
Find the first integer, −3, on the number line.

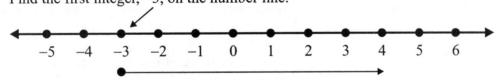

Step 2:   (−3) + 7 The second integer in the problem, +7, tells us the direction to go,
positive (toward positive numbers), and how far, 7 places. (−3) + 7 = 4

EXAMPLE 2: Add: (−2) + (−3)

Step 1:   Find the first integer, (−2), on the number line.

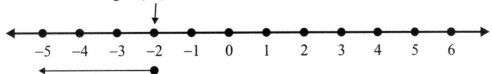

Step 2:   (−2) + (−3) The second integer in the problem, (−3), tells us the direction to go,
negative (toward the negative numbers), and how far, 3 places. (−2) + (−3) = (−5)

175

Solve the problems below using this number line.

$$\xleftarrow{\quad} \overset{\bullet}{-8} \; \overset{\bullet}{-7} \; \overset{\bullet}{-6} \; \overset{\bullet}{-5} \; \overset{\bullet}{-4} \; \overset{\bullet}{-3} \; \overset{\bullet}{-2} \; \overset{\bullet}{-1} \; \overset{\bullet}{0} \; \overset{\bullet}{1} \; \overset{\bullet}{2} \; \overset{\bullet}{3} \; \overset{\bullet}{4} \; \overset{\bullet}{5} \; \overset{\bullet}{6} \; \overset{\bullet}{7} \; \overset{\bullet}{8} \xrightarrow{\quad}$$

1. $2 + (-3) =$ _____
2. $4 + (-2) =$ _____
3. $(-3) + 7 =$ _____
4. $(-4) + 4 =$ _____
5. $(-1) + 5 =$ _____
6. $(-1) + (-4) =$ _____
7. $3 + 2 =$ _____
8. $(-5) + 8 =$ _____

9. $3 + (-7) =$ _____
10. $(-2) + (-2) =$ _____
11. $6 + (-7) =$ _____
12. $2 + (-5) =$ _____
13. $(-5) + 3 =$ _____
14. $(-6) + 7 =$ _____
15. $(-3) + (-3) =$ _____
16. $(-8) + 6 =$ _____

17. $(-2) + 6 =$ _____
18. $(-4) + 8 =$ _____
19. $(-7) + 4 =$ _____
20. $(-5) + 8 =$ _____
21. $(-2) + (-2) =$ _____
22. $8 + (-6) =$ _____
23. $5 + (-3) =$ _____
24. $1 + (-8) =$ _____

RULES FOR ADDING INTEGERS WITH THE SAME SIGNS

To add integers without using the number line, use these simple rules:

> 1. **Add the numbers together.**
> 2. **Give the answer the same sign.**

EXAMPLE 1: $(-2) + (-5) =$ _____   Both integers are negative. To find the answer, add the numbers together $(2 + 5)$, and give the answer a negative sign.

$(-2) + (-5) = (-7)$

EXAMPLE 2: $3 + 4 =$ _____   Both integers are positive so the answer is positive.

$3 + 4 = 7$ **NOTE:** Sometimes positive signs are placed in front of positive numbers. For example $3 + 4 = 7$ may be written $(+3) + (+4) = +7$. Positive signs in front of positive numbers are optional. If a number has no sign, it is considered positive.

Solve the problems below using the rules for adding integers with the same signs.

1. $(-18) + (-4) =$ _____
2. $(-12) + (-3) =$ _____
3. $(-2) + (-7) =$ _____
4. $(+22) + (+11) =$ _____
5. $(-7) + (-6) =$ _____

6. $(-9) + (-8) =$ _____
7. $8 + 4 =$ _____
8. $(-4) + (-7) =$ _____
9. $(-15) + (-5) =$ _____
10. $(+7) + (+4) =$ _____

11. $(-7) + (-20) =$ _____
12. $(-18) + (-16) =$ _____
13. $25 + 32 =$ _____
14. $(-15) + (-3) =$ _____
15. $(9) + (9) =$ _____

176

RULES FOR ADDING INTEGERS WITH OPPOSITE SIGNS

> 1. Ignore the signs and find the difference.
> 2. Give the answer the sign of the larger number.

EXAMPLE 1: $(-4) + 6 = \underline{\hspace{1cm}}$

$(-4) + 6 = 2$

To find the difference, take the larger number minus the smaller number. $6 - 4 = 2$. Looking back at the original problem, the larger number, 6, is positive, so the answer is positive.

EXAMPLE 2: $3 + (-7) = \underline{\hspace{1cm}}$

$3 + (-7) = (-4)$

Find the difference. $7 - 3 = 4$. Looking at the problem, the larger number, 7, is a negative number, so the answer is negative.

Solve the problems below using the rules of adding integers with opposite signs.

1. $(-4) + 8 = \underline{\hspace{1cm}}$
2. $-10 + 12 = \underline{\hspace{1cm}}$
3. $9 + (-3) = \underline{\hspace{1cm}}$
4. $(+3) + (-3) = \underline{\hspace{1cm}}$
5. $5 + (-2) = \underline{\hspace{1cm}}$
6. $(-18) + 9 = \underline{\hspace{1cm}}$
7. $25 + (-30) = \underline{\hspace{1cm}}$

8. $+8 + (-7) = \underline{\hspace{1cm}}$
9. $(-5) + (+12) = \underline{\hspace{1cm}}$
10. $-14 + (+7) = \underline{\hspace{1cm}}$
11. $7 + (-8) = \underline{\hspace{1cm}}$
12. $(-30) + 15 = \underline{\hspace{1cm}}$
13. $100 + (-65) = \underline{\hspace{1cm}}$
14. $85 + (-14) = \underline{\hspace{1cm}}$

15. $6 + (-12) = \underline{\hspace{1cm}}$
16. $(-11) + 1 = \underline{\hspace{1cm}}$
17. $3 + (-13) = \underline{\hspace{1cm}}$
18. $(-12) + 8 = \underline{\hspace{1cm}}$
19. $52 + (-9) = \underline{\hspace{1cm}}$
20. $(-39) + 8 = \underline{\hspace{1cm}}$
21. $(+14) + (-16) = \underline{\hspace{1cm}}$

Solve the mixed addition problems below using the rules for adding integers.

22. $-7 + 8 = \underline{\hspace{1cm}}$
23. $5 + 6 = \underline{\hspace{1cm}}$
24. $(-2) + (-6) = \underline{\hspace{1cm}}$
25. $3 + (-5) = \underline{\hspace{1cm}}$
26. $(-7) + (-9) = \underline{\hspace{1cm}}$
27. $14 + 9 = \underline{\hspace{1cm}}$
28. $(-15) + 6 = \underline{\hspace{1cm}}$

29. $8 + (-5) = \underline{\hspace{1cm}}$
30. $(-6) + 13 = \underline{\hspace{1cm}}$
31. $(-9) + (-12) = \underline{\hspace{1cm}}$
32. $(-7) + (+12) = \underline{\hspace{1cm}}$
33. $+8 + (-9) = \underline{\hspace{1cm}}$
34. $(-13) + (-18) = \underline{\hspace{1cm}}$
35. $46 + (-52) = \underline{\hspace{1cm}}$

36. $(-7) + (+10) = \underline{\hspace{1cm}}$
37. $(+4) + 11 = \underline{\hspace{1cm}}$
38. $11 + 6 = \underline{\hspace{1cm}}$
39. $-4 + (-10) = \underline{\hspace{1cm}}$
40. $(+6) + (+2) = \underline{\hspace{1cm}}$
41. $1 + (-17) = \underline{\hspace{1cm}}$
42. $(-42) + 6 = \underline{\hspace{1cm}}$

RULES FOR SUBTRACTING INTEGERS

To subtract integers, the easiest way is to change the problem to an addition problem and follow the rules you already know.

> 1. Change the subtraction sign to addition.
> 2. Change the sign of the second number to the opposite sign.

EXAMPLE 1: $-6 - (-2) =$ _____

$(-6) + 2 = (-4)$

Change the subtraction sign to addition and -2 to 2. $-6 - (-2) = (-6) + 2$

EXAMPLE 2: $5 - 6 =$ _____

$5 + (-6) = (-1)$

Change the subtraction sign to addition and 6 to -6. $5 - 6 = 5 + (-6)$

Solve the problems using the rules above.

1. $(-3) - 8 =$ _____
2. $5 - (-9) =$ _____
3. $8 - (-5) =$ _____
4. $(-2) - (-6) =$ _____
5. $8 - (-9) =$ _____
6. $(-4) - (-1) =$ _____

7. $(-5) - (-13) =$ _____
8. $6 - (-7) =$ _____
9. $8 - (-6) =$ _____
10. $(-2) - (-2) =$ _____
11. $(-3) - 7 =$ _____
12. $(-4) - 8 =$ _____

13. $(-7) - 4 =$ _____
14. $1 - (-9) =$ _____
15. $(-5) - 12 =$ _____
16. $(-1) - 9 =$ _____
17. $6 - (-7) =$ _____
18. $(-8) - (-12) =$ _____

Solve the addition and subtraction problems below.

19. $4 - (-2) =$ _____
20. $(-3) + 7 =$ _____
21. $(-4) + 14 =$ _____
22. $(-1) - 5 =$ _____
23. $(-1) + (-4) =$ _____
24. $(-12) + (-2) =$ _____
25. $0 - (-6) =$ _____
26. $2 - (-5) =$ _____

27. $(-5) + 3 =$ _____
28. $(-6) + 7 =$ _____
29. $(-4) + 8 =$ _____
30. $(-4) - 11 =$ _____
31. $(-5) + 8 =$ _____
32. $(-3) - (-3) =$ _____
33. $(-8) + 9 =$ _____
34. $0 + (-10) =$ _____

35. $30 + (-15) =$ _____
36. $-40 - (-5) =$ _____
37. $25 - 50 =$ _____
38. $-13 + 12 =$ _____
39. $(-21) - (-1) =$ _____
40. $62 - (-3) =$ _____
41. $(-16) + (-2) =$ _____
42. $(-25) + 5 =$ _____

MULTIPLYING INTEGERS

You are probably used to seeing multiplication written with a "×" sign, but multiplication can be written two other ways. A "·" between numbers means the same as "×", and parentheses ( ) around a number without a "×" or a "·" also means to multiply.

EXAMPLE: $2 \times 3 = 6$ or $2 \cdot 3 = 6$ or $(2)(3) = 6$

All of these mean the same thing.

DIVIDING INTEGERS

Division is commonly indicated two ways: with a "÷" or in the form of a fraction.

EXAMPLE: $6 \div 3 = 2$ means the same thing as $\dfrac{6}{3} = 2$

RULES FOR MULTIPLYING AND DIVIDING INTEGERS

> 1.   **If the numbers have the same sign, the answer is positive.**
> 2.   **If the numbers have different signs, the answer is negative.**

EXAMPLES: $6 \times 8 = 48$ $(-6) \times 8 = (-48)$ $(-6) \times (-8) = 48$

$48 \div 6 = 8$ $(-48) \div 6 = (-8)$ $(-48) \div (-6) = 8$

Solve the problems below using the rules for multiplying and dividing integers with opposite signs.

1. $(-4) \div 2 =$ _____

2. $12 \div (-3) =$ _____

3. $\dfrac{(-14)}{(-2)} =$ _____

4. $-15 \div 3 =$ _____

5. $(-3) \times (-7) =$ _____

6. $(-1) \cdot (5) =$ _____

7. $-1 \times (-4) =$ _____

8. $(3)(2) =$ _____

9. $2(-5) =$ _____

10. $3 \times (-7) =$ _____

11. $(-12) \cdot (-2) =$ _____

12. $\dfrac{(-18)}{(-6)} =$ _____

13. $21 \div (-7) =$ _____

14. $-5 \times 3 =$ _____

15. $(-6)(7) =$ _____

16. $\dfrac{(-3)}{(-3)} =$ _____

17. $(-5) \times 8 =$ _____

18. $\dfrac{-12}{6} =$ _____

19. $8(-4) =$ _____

20. $1 \cdot (-8) =$ _____

21. $(-7) \cdot (-4) =$ _____

22. $(-2) \div (-2) =$ _____

23. $\dfrac{18}{(-6)} =$ _____

24. $5(-3) =$ _____

MIXED INTEGER PRACTICE

1. $(-6) + 13 =$ _____
2. $(-3) + (-9) =$ _____
3. $(-4) \times 4 =$ _____
4. $(-18) \div 3 =$ _____
5. $(-1) - 5 =$ _____
6. $(-1) \times (-4) =$ _____
7. $3 + (-5) =$ _____
8. $6 + (-5) =$ _____
9. $(-9) - (-12) =$ _____

10. $2 + (-5) =$ _____
11. $\dfrac{(-24)}{(-6)} =$ _____
12. $(-5) + 3 =$ _____
13. $(-6) - 7 =$ _____
14. $(-33) \div (-11) =$ _____
15. $(-21)(-3) =$ _____
16. $(-7) + (-14) =$ _____
17. $(-5) - 8 =$ _____
18. $1(-8) =$ _____

19. $(-2) \cdot (-2) =$ _____
20. $8 + (-6) =$ _____
21. $\dfrac{-14}{7} =$ _____
22. $(+7) \cdot (-2) =$ _____
23. $(10)(4) =$ _____
24. $24 \div (-4) =$ _____
25. $6(-5) =$ _____
26. $\dfrac{12}{(-3)} =$ _____
27. $36 \div 12 =$ _____

INTEGER WORD PROBLEMS

Solve the following word problems.

1. If it is 2° outside and the temperature will drop 15° tonight, how cold will it get?

2. Karen has $12 left in her checking account. She needs to write a check for $44. What will her balance be if she writes the check?

3. It is −24° tonight, but the weather reporter predicted it would be 16° warmer tomorrow. What will the temperature be tomorrow?

4. The average temperature of the earth's stratosphere is −70°F. The average temperature on the earth's surface is 57°F. How much warmer is the average surface of the earth than the stratosphere?

5. The elevation of the Dead Sea is −1,286 feet. (The Dead Sea is below sea level.) Mt. McKinley has an elevation of 20,320 feet. What is the difference in the elevation between the Dead Sea and Mt. McKinley?

6. A submarine dives 462 feet beneath the surface of the ocean. It then climbs up 257 feet. What depth is the submarine now?

7. Eratosthenes was born about 274 B.C. Sir Isaac Newton was born in 1642 A.D. About how many years apart were they born?

8. The average daily low temperature in International Falls, Minnesota during the month of January is −9°F. The average high is 14°F. What is the temperature difference between the average low and the average high?

ORDER OF OPERATIONS

In long math problems with $+, -, \times, \div, (\ )$, and exponents in them, you have to know what to do first. Without following the same rules, you could get different answers. If you will memorize the silly sentence, "Please Excuse My Dear Aunt Sally," you can easily memorize what order you must follow.

<u>P</u>lease   **"P"** stands for parentheses. You must get rid of parentheses first.
Examples: $3(1 + 4) = 3 \times 5 = 15$
                 $6(10 - 6) = 6 \times 4 = 24$

<u>E</u>xcuse   **"E"** stands for exponents. You must eliminate exponents next.
Example: $4^2 = 4 \times 4 = 16$

<u>M</u>y <u>D</u>ear   **"M"** stands for multiply. **"D"** stands for divide. Start on the left of the equation, and perform all multiplications and divisions in the order in which they appear.

<u>A</u>unt <u>S</u>ally   **"A"** stands for add. **"S"** stands for subtract. Start on the left, and perform all additions and subtractions in the order they appear.

EXAMPLE: $12 \div 2(6 - 3) + 3^2 - 1$

Please	Eliminate **parentheses**. $6 - 3 = 3$ so now we have	$12 \div 2 \times 3 + 3^2 - 1$
Excuse	Eliminate **exponents**. $3^2 = 9$ so now we have	$12 \div 2 \times 3 + 9 - 1$
My Dear	**Multiply** and **divide** next in order from left to right.	$12 \div 2 = 6$ then $6 \times 3 = 18$
Aunt Sally	Last, we **add** and **subtract** in order from left to right.	$18 + 9 - 1 = 26$

Simplify the following problems.

1. $6 + 9 \times 2 - 4 =$ _____

2. $3(4 + 2) - 6^2 =$ _____

3. $3(6 - 3) - 2^3 =$ _____

4. $49 \div 7 - 3 \times 3 =$ _____

5. $10 \times 4 - (7 - 2) =$ _____

6. $2 \times 3 \div 6 \times 4 =$ _____

7. $50 - 8(4 + 2) =$ _____

8. $7 + 8(14 - 6) \div 4 =$ _____

9. $(2 + 8 - 12) \times 4 =$ _____

10. $4(8 - 13) \times 4 =$ _____

11. $8 + 4^2 \times 2 - 6 =$ _____

12. $3^2(4 + 6) + 3 =$ _____

13. $(12 - 6) + 27 \div 3^2 =$ _____

14. $82^0 - 1 + 4 \div 2^2 =$ _____

15. $1 - (2 - 3) + 8 =$ _____

16. $12 - 4(7 - 2) =$ _____

17. $18 \div (6 + 3) - 12 =$ _____

18. $10^2 + 3^3 - 2 \times 3 =$ _____

19. $4^2 + (7 + 2) \div 3 =$ _____

20. $7 \times 4 - 9 \div 3 =$ _____

CHAPTER 12 REVIEW

Simplify the following problems.

1. $(-9) \times (-10) =$ _____
2. $12 + (-22) =$ _____
3. $10 - (-13) =$ _____
4. $12 \div (-3) =$ _____
5. $|-5| + |4| =$ _____
6. $-6 - 5 =$ _____
7. $(-7) \cdot (6) =$ _____
8. $-|9| - |-2| =$ _____
9. $4 - 9 =$ _____
10. $\frac{(-25)}{(-5)} =$ _____
11. $(-13) + (-4) =$ _____
12. $(-10)(-6) =$ _____

13. $|-10| + |-2| =$ _____
14. $(-4)(-22) =$ _____
15. $3 + (-9) =$ _____
16. $\frac{(-16)}{4} =$ _____
17. $(6)(-11) =$ _____
18. $-4 + (-10) =$ _____
19. $(-24) \div (-6) =$ _____
20. $13 - 18 =$ _____
21. $14 + (-20) =$ _____
22. $\frac{45}{(-9)} =$ _____
23. $|-7| + |-5| =$ _____
24. $-3 - (-3) =$ _____

25. $(-1) \times 12 =$ _____
26. $-7 + (-27) =$ _____
27. $9 - (-4) =$ _____
28. $(12)(-3) =$ _____
29. $\frac{(-60)}{(-12)} =$ _____
30. $13 - 27 =$ _____
31. $(-15) \times (-2) =$ _____
32. $-|4| - |-8| =$ _____
33. $(-19) + 8 =$ _____
34. $-13 - (-13) =$ _____
35. $(-7) \cdot (-8) =$ _____
36. $36 \div (-6) =$ _____

Simplify the following problems using the correct order of operations.

37. $2^3 + (2^2)(8 - 12) =$ _____
38. $20 \div (-2 - 8) + 2 =$ _____
39. $14 + (7)(3 - 5) \div 2 =$ _____
40. $8 - 7^2 + (3 - 12) =$ _____
41. $(18 - 5) \times (2 - 3) - 10 =$ _____

42. $24 - (6^2 - 6) \div 5 =$ _____
43. $3^3 + (7)(9 - 5) =$ _____
44. $-10(1 - 9) \div (-20) + 1 =$ _____
45. $42 \div (12 - 5) - 2 =$ _____
46. $2 + 5^2 \div (15 - 20) =$ _____

Solve the following word problems.

47. Aristotle, an ancient Greek philosopher, was born in 384 B.C. Roger Bacon, a famous English philosopher, was born about 1214 A.D. About how many years apart were they born?

48. Echo River flows through a deep cavern in Mammoth Cave in southwest Kentucky 360 feet below the surface. The highest peak of the Appalachian Mountains is Mount Mitchell with a height of 6684 feet. What is the difference in elevation between Mammoth Cave's Echo River and the top of Mount Mitchell?

Chapter 13

PLANE GEOMETRY

The following terms are important for understanding the concepts that are presented in this chapter. Most of you have already been introduced to these terms. Rather than defining them in words, they are presented here by example as a refresher.

LINE SEGMENTS

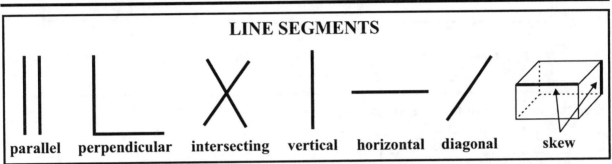

parallel   perpendicular   intersecting   vertical   horizontal   diagonal   skew

TRIANGLES

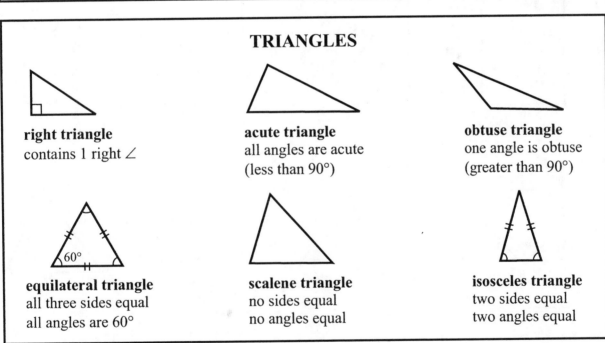

right triangle
contains 1 right ∠

acute triangle
all angles are acute
(less than 90°)

obtuse triangle
one angle is obtuse
(greater than 90°)

equilateral triangle
all three sides equal
all angles are 60°

scalene triangle
no sides equal
no angles equal

isosceles triangle
two sides equal
two angles equal

POLYGONS

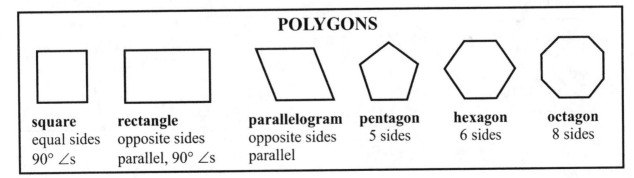

square
equal sides
90° ∠s

rectangle
opposite sides
parallel, 90° ∠s

parallelogram
opposite sides
parallel

pentagon
5 sides

hexagon
6 sides

octagon
8 sides

PERIMETER

The **perimeter** is the distance around a polygon. To find the perimeter, add the lengths of the sides.

EXAMPLES:

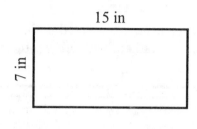

$P = 7 + 15 + 7 + 15$
$P = 44$ in

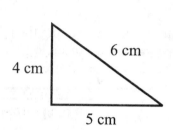

$P = 4 + 6 + 5$
$P = 15$ cm

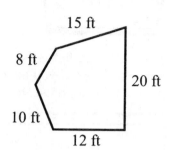

$P = 8 + 15 + 20 + 12 + 10$
$P = 65$ ft

Find the perimeter of the following polygons.

1.

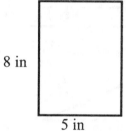

_____

4.

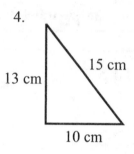

_____

7.

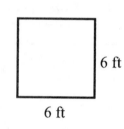

_____

10.

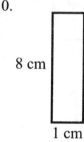

_____

2.

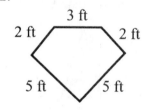

_____

5.

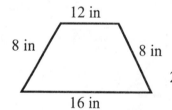

_____

8.

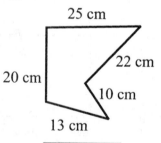

_____

11.

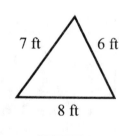

_____

3.

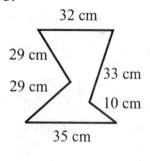

_____

6.

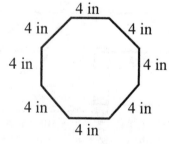

_____

9.

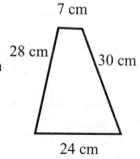

_____

12.

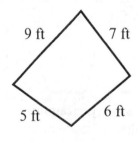

_____

AREA OF SQUARES AND RECTANGLES

Area - area is always expressed in square units such as in^2, cm^2, ft^2, and m^2.

The area, (A), of squares and rectangles equals length (ℓ) times width (w). $A = \ell\, w$

EXAMPLE:

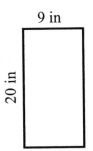

4 cm

4 cm

$A = \ell\, w$
$A = 4 \times 4$
$A = 16\ \mathbf{cm}^2$

If a square has an area of 16 cm^2, it means that it will take 16 squares that are 1 cm on each side to cover the area of a square that is 4 cm on each side.

Find the area of the following squares and rectangles using the formula $A = \ell w$.

1.
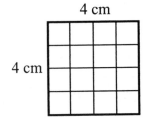

10 ft

10 ft

_____

2.
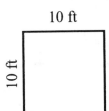

5 cm

2 cm

_____

3.
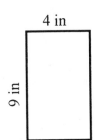

4 in

9 in

_____

4.
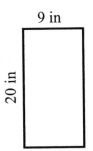

9 in

20 in

_____

5.
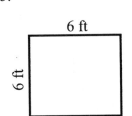

6 ft

6 ft

_____

6.
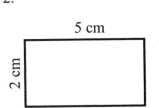

10 cm

5 cm

_____

7.
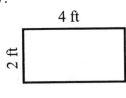

4 ft

2 ft

_____

8.
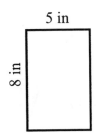

5 in

8 in

_____

9.
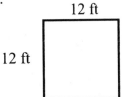

12 ft

12 ft

_____

10.
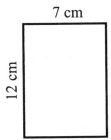

7 cm

12 cm

_____

11.
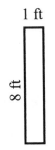

1 ft

8 ft

_____

12.
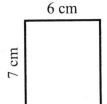

6 cm

7 cm

_____

185

AREA OF TRIANGLES

EXAMPLE: Find the area of the following triangle.

The formula for the area of a triangle is:

$A = \frac{1}{2} \times b \times h$

A = area
b = base
h = height

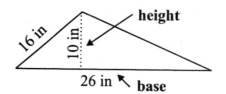

Step 1:   Insert measurements from the triangle into the formula: $A = \frac{1}{2} \times 26 \times 10$

Step 2:   Cancel and multiply. $A = \frac{1}{\cancel{2}} \times \frac{\cancel{26}^{13}}{1} \times \frac{10}{1} = 130 \text{ in}^2$

Note: Area is always expressed in square units such as in^2, ft^2, cm^2, or m^2.

Find the area of the following triangles. Remember to include units.

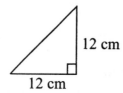
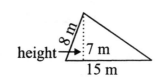
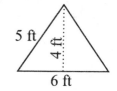

1. _____ in^2 4. _____ cm^2 7. _____ m^2 10. _____ ft^2

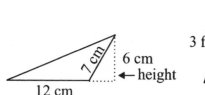
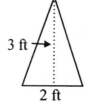
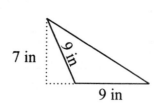
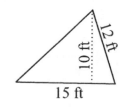

2. _____ cm^2 5. _____ ft^2 8. _____ in^2 11. _____ ft^2

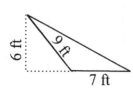
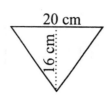
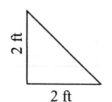
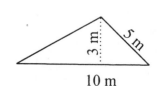

3. _____ ft^2 6. _____ cm^2 9. _____ ft^2 12. _____ m^2

AREA OF TRAPEZOIDS AND PARALLELOGRAMS

EXAMPLE 1: Find the area of the following parallelogram.

The formula for the area of a parallelogram is: $A = bh$

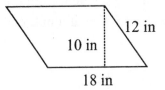

    A = **area**
    b = **base**
    h = **height**

Step 1: Insert measurements from the parallelogram into the formula: $A = 18 \times 10$

Step 2: Multiply. $18 \times 10 = 180$ in^2

EXAMPLE 2: Find the area of the following trapezoid.

The formula for the area of a trapezoid is: $A = \frac{1}{2} h\,(b_1 + b_2)$. A trapezoid has two bases that are parallel to each other. When you add the length of the two bases together and then multiply by $\frac{1}{2}$, you find their average length.

    A = **area**
    b = **base**
    h = **height**

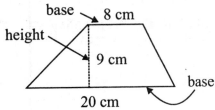

Insert measurements from the trapezoid into the formula and solve:
$\frac{1}{2} \times 9\ (8 + 20) = 126$ cm^2

Find the area of the following parallelograms and trapezoids.

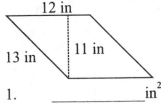

1. _____ in^2

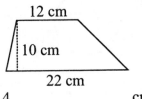

4. _____ cm^2

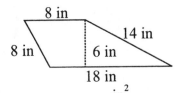

7. _____ in^2

2. _____ in^2

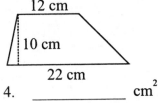

5. _____ in^2

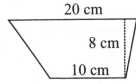

8. _____ cm^2

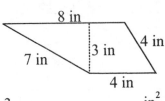

3. _____ in^2

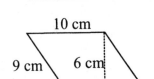

6. _____ cm^2

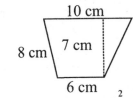

9. _____ cm^2

CIRCUMFERENCE

Circumference, _C_ - the distance around the outside of a circle
Diameter, _d_ - a line segment passing through the center of a circle from one side to the other
Radius, _r_ - a line segment from the center of a circle to the edge of the circle
Pi, π - the ratio of the circumference of a circle to its diameter $\pi = 3.14$ or $\pi = \dfrac{22}{7}$

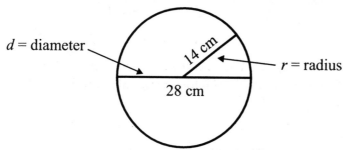

The formula for the circumference of a circle is $C = 2\pi r$ or $C = \pi d$ (The formulas are equal because the diameter is equal to twice the radius, $d = 2r$.)

EXAMPLE:

Find the circumference of the circle above.

$C = \pi d$ Use $= 3.14$
$C = 3.14 \times 28$
$C = 87.92$ cm

EXAMPLE:

Find the circumference of the circle above.

$C = 2\pi r$
$C = 2 \times 3.14 \times 14$
$C = 87.92$ cm

Use the formulas given above to find the circumference of the following circles.
Use π = 3.14.

1.
   8 in

2.
   14 ft

3.
   2 cm

4.
   6 m

5.
   8 ft

$C = $ _____ $C = $ _____ $C = $ _____ $C = $ _____ $C = $ _____

Use the formulas given above to find the circumference of the following circles.
Use π = $\dfrac{22}{7}$.

6.
   3 ft

7.
   12 in

8.
   6 m

9.
   5 cm

10.
    16 in

$C = $ _____ $C = $ _____ $C = $ _____ $C = $ _____ $C = $ _____

AREA OF A CIRCLE

The formula for area of a circle is $A = \pi r^2$. The area is how many square units of measure would fit inside a circle.

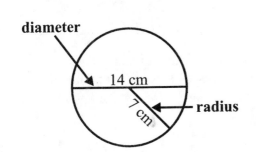

diameter

14 cm

7 cm

radius

$\pi = \dfrac{22}{7}$ or $\pi = 3.14$

EXAMPLE: Find the area of the circle using both values for π.

Let $\pi = \dfrac{22}{7}$

$A = \pi r^2$

$A = \dfrac{22}{7} \times 7^2$

$A = \dfrac{22}{\cancel{7}} \times \dfrac{\cancel{49}^{7}}{1} = 154 \text{ cm}^2$

Let $\pi = 3.14$

$A = \pi r^2$

$A = 3.14 \times 7^2$

$A = 3.14 \times 49 = 153.86 \text{ cm}^2$

Find the area of the following circles. Remember to include units.

$\pi = 3.14$ $\pi = \dfrac{22}{7}$

1.
   5 in

   $A = \underline{\hspace{1.5cm}}$ $A = \underline{\hspace{1.5cm}}$

2.
   16 ft

   $A = \underline{\hspace{1.5cm}}$ $A = \underline{\hspace{1.5cm}}$

3.
   8 cm

   $A = \underline{\hspace{1.5cm}}$ $A = \underline{\hspace{1.5cm}}$

4.
   3 m

   $A = \underline{\hspace{1.5cm}}$ $A = \underline{\hspace{1.5cm}}$

Fill in the chart below. Include appropriate units.

	Radius	Diameter	Area $\pi = 3.14$	Area $\pi = \dfrac{22}{7}$
5.	9 ft			
6.		4 in		
7.	8 cm			
8.		20 ft		
9.	14 m			
10.		18 cm		
11.	12 ft			
12.		6 in		

AREA AND CIRCUMFERENCE MIXED PRACTICE

1. Find the area of a circle with a diameter of 12 centimeters. Use $\pi = 3.14$.

    _____

2. Which circle has a larger circumference, a circle with a radius of 9 feet or a circle with a diameter of 5 feet?

    _____

3. Find the circumference of a circle that is 14 centimeters in diameter. Use $\pi = \frac{22}{7}$.

    _____

4. Paula has a round garden. The diameter is 14 feet. What is the area of the garden? Use $\pi = 3.14$.

    _____

Find the area and circumference of each circle. Use $\pi = 3.14$.

5.
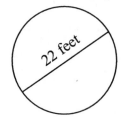
22 feet

$A = $ _____

$C = $ _____

6.
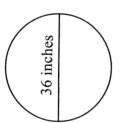
36 inches

$A = $ _____

$C = $ _____

7.
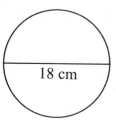
18 cm

$A = $ _____

$C = $ _____

Find the area and the circumference of each circle. Use $\pi = \frac{22}{7}$.

8.
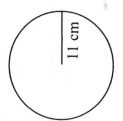
11 cm

$A = $ _____

$C = $ _____

9.
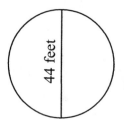
44 feet

$A = $ _____

$C = $ _____

10.
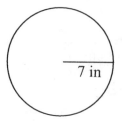
7 in

$A = $ _____

$C = $ _____

190

TWO-STEP AREA PROBLEMS

Solving the problems below will require two steps. You will need to find the area of two figures, and then either add or subtract the two areas to find the answer. **Carefully read the EXAMPLES below.**

EXAMPLE

Find the area of the living room below.

Figure 1

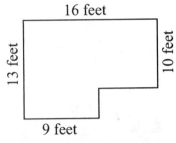

Step 1: Complete the rectangle as in Figure 2, and figure the area as if it were a complete rectangle.

Figure 2

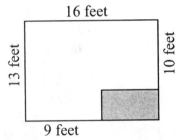

$A = \text{length} \times \text{width}$
$A = 16 \times 13$
$A = 208 \text{ ft}^2$

Step 2: Figure the area of the shaded part.

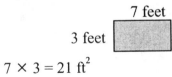

7 feet

3 feet

$7 \times 3 = 21 \text{ ft}^2$

Step 3: Subtract the area of the shaded part from the area of the complete rectangle.

$208 - 21 = 187 \text{ ft}^2$

EXAMPLE

Find the area of the shaded sidewalk.

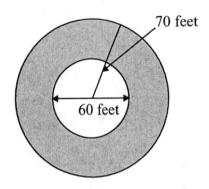

Step 1: Find the area of the outside circle.
$\pi = 3.14$
$A = 3.14 \times 70 \times 70$
$A = 15,386 \text{ ft}^2$

Step 2: Find the area of the inside circle.
$\pi = 3.14$
$A = 3.14 \times 30 \times 30$
$A = 2,826 \text{ ft}^2$

Step 3: Subtract the area of the inside circle from the area of the outside circle.

$15,386 - 2,826 = 12,560 \text{ ft}^2$

Find the area of the following figures.

1.

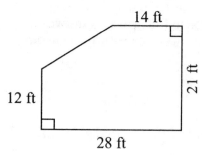

14 ft

12 ft

21 ft

28 ft

_____

2.

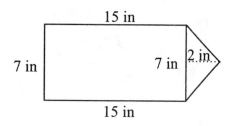

15 in

7 in   7 in   2 in

15 in

_____

3.   What is the area of the shaded circle? Use π = 3.14 and round the answer to the nearest whole number.

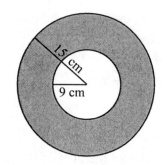

15 cm

9 cm

_____

4.

1 ft

5 ft

18 ft

4 ft

_____

5.   What is the area of the rectangle that is shaded? Use π = 3.14 and round to the nearest whole number.

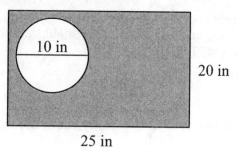

10 in

20 in

25 in

_____

6.   What is the area of the shaded part?

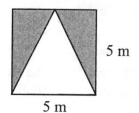

5 m

5 m

_____

7.   What is the area of the shaded part?

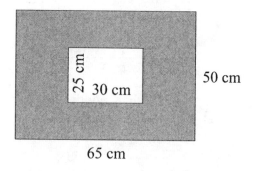

25 cm   30 cm   50 cm

65 cm

_____

8. 6 m

24 m

12 m

12 m

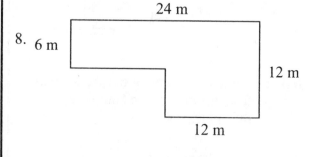

_____

MEASURING TO FIND PERIMETER AND AREA

Using a ruler, measure the perimeter of the following figures, and calculate the area. Be careful to measure in the correct unit. To calculate the area of some of the figures, you will have to use the two-step approach.

1.

$P =$ _____ cm

$A =$ _____ cm^2

4.

$P =$ _____ in

$A =$ _____ in^2

7.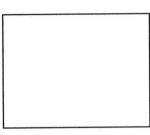

$P =$ _____ cm

$A =$ _____ cm^2

2.

$P =$ _____ in

$A =$ _____ in^2

5.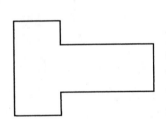

$P =$ _____ in

$A =$ _____ in^2

8.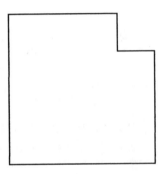

$P =$ _____ cm

$A =$ _____ cm^2

3.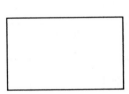

$P =$ _____ in

$A =$ _____ in^2

6.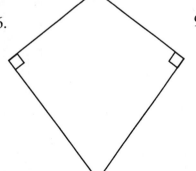

$P =$ _____ cm

$A =$ _____ cm^2

9.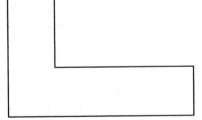

$P =$ _____ in

$A =$ _____ in^2

ESTIMATING AREA

To estimate the area of the an object using a grid, follow the steps below.

Step 1: Count the number of whole squares that are shaded.   6

Step 2: Count the number of squares that are at least half shaded.   2

Step 3: Find the sum of the two numbers. $6 + 2 = 8$

The area is about 8 cm^2.

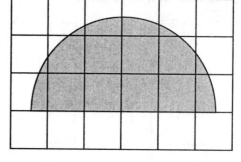

Each square is equal to 1 cm^2.

Estimate the area of each of the shaded figures below to the nearest square centimeter.

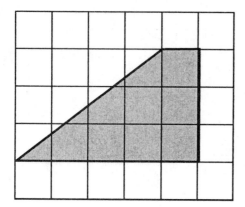

1.   Each square is equal to 1 cm^2.
The area is about _____ cm^2.

3.   Each square is equal to 1 cm^2.
The area is about _____ cm^2.

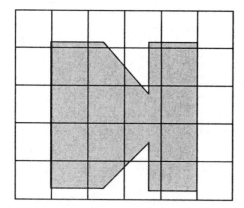

2.   Each square is equal to 1 cm^2. The area is about _____ cm^2.

4.   Each square is equal to 1 cm^2.
The area is about _____ cm^2.

Estimate the area of each of the shaded figures to the nearest square yard or square mile.

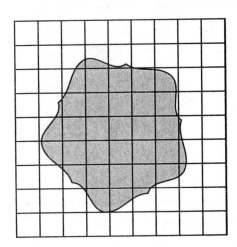

1. 1 square = 1 square mile
    The area is about _____
    square miles.

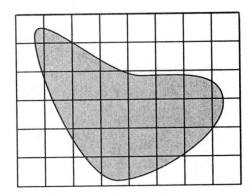

2. 1 square = 1 square mile
    The area is about _____
    square miles.

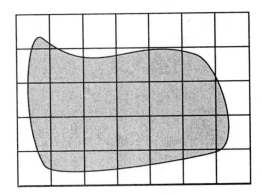

3. 1 square = 1 square mile
    The area is about _____
    square miles.

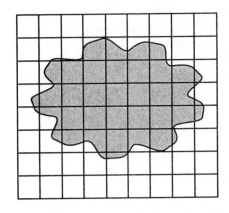

4. 1 square = 1 square yard
    The area is about _____
    square yards.

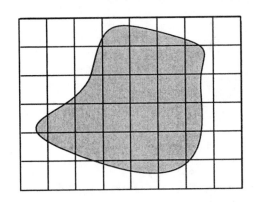

5. 1 square = 1 square yard
    The area is about _____
    square yards.

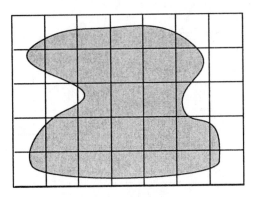

6. 1 square = 1 square yard
    The area is about _____
    square yards.

SIMILAR AND CONGRUENT

Similar figures have the same shape but are two different sizes. Their corresponding sides are proportional. **Congruent figures** are exactly alike in size, and shape and their corresponding sides are equal. See the examples below.

SIMILAR   **CONGRUENT**

Directions: Label each pair of figures below as either S if they are similar, C if they are congruent, or N if they are neither.

1. _____

2. _____

3. _____

4. _____

5. _____

6. _____

7. _____

8. _____

9. _____

10. _____

11. _____

12. _____

CARTESIAN COORDINATES

A number line allows you to graph points with only one value. A **Cartesian coordinate plane** allows you to graph points with two values. A Cartesian coordinate plane is made up of two number lines. The horizontal number line is called the **x-axis**, and the vertical number line is called the **y-axis**. The point where the x and y axes intersect is called the **origin**. The x and y axes separate the Cartesian coordinate plane into four quadrants that are labeled I, II, III, and IV. The quadrants are labeled and explained on the graph below. Each point graphed on the plane is designated by an **ordered pair** of coordinates. For example, $(2, -1)$ is an ordered pair of coordinates designated by **point B** on the plane below. The first number, 2, tells you to go over positive two on the x-axis. The -1 tells you to then go down negative one on the y-axis.

Remember: The first number always tells you how far to go right or left of 0, and the second number always tells you how far to go up or down from 0.

Quadrant II:
The x-coordinate is negative, and the y-coordinate is positive $(-, +)$.

Quadrant III:
Both coordinates in the ordered pair are negative $(-, -)$.

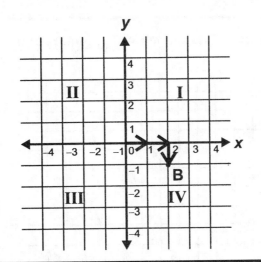

Quadrant I:
Both coordinates in the ordered pair are positive $(+, +)$.

Quadrant IV:
The x-coordinate is positive, and the y-coordinate is negative $(+, -)$.

Plot and label the following points on the Cartesian coordinate plane provided.

A.	$(2, 4)$	G.	$(-2, 5)$	M.	$(5, 5)$
B.	$(-1, 5)$	H.	$(5, -1)$	N.	$(-2, -2)$
C.	$(3, -4)$	I.	$(4, -4)$	O.	$(0, 0)$
D.	$(-5, -2)$	J.	$(5, 2)$	P.	$(0, 4)$
E.	$(5, 3)$	K.	$(-1, -1)$	Q.	$(2, 0)$
F.	$(-3, -5)$	L.	$(3, -3)$	R.	$(-4, 0)$

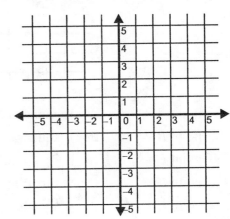

IDENTIFYING ORDERED PAIRS

When identifying **ordered pairs**, count how far left or right of 0 to find the *x*-coordinate and then how far up or down from 0 to find the *y*-coordinate.

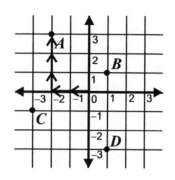

Point A: Left (negative) two and up (positive) three = (–2, 3) in quadrant II

Point B: Right (positive) one and up (positive) one = (1, 1) in quadrant I

Point C: Left (negative) three and down (negative) one = (–3, –1) in quadrant III

Point D: Right (positive) one and down (negative) three = (1, –3) in quadrant IV

Fill in the ordered pair for each point, and tell which quadrant it is in.

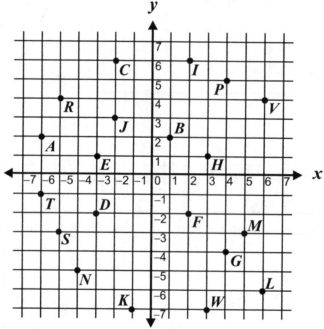

1. point A =   (,   )   quadrant _____
2. point B =   (,   )   quadrant _____
3. point C =   (,   )   quadrant _____
4. point D =   (,   )   quadrant _____
5. point E =   (,   )   quadrant _____
6. point F =   (,   )   quadrant _____
7. point G =   (,   )   quadrant _____
8. point H =   (,   )   quadrant _____
9. point I =   (,   )   quadrant _____
10. point J =   (,   )   quadrant _____

11. point K =   (,   )   quadrant _____
12. point L =   (,   )   quadrant _____
13. point M =   (,   )   quadrant _____
14. point N =   (,   )   quadrant _____
15. point P =   (,   )   quadrant _____
16. point R =   (,   )   quadrant _____
17. point S =   (,   )   quadrant _____
18. point T =   (,   )   quadrant _____
19. point V =   (,   )   quadrant _____
20. point W =   (,   )   quadrant _____

198

Sometimes, points on a coordinate plane fall on the x or y axis. If a point falls on the x-axis, then the second number of the ordered pair is 0. If a point falls on the y-axis, the first number of the ordered pair is 0.

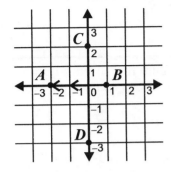

Point A: Left (negative) two and up zero = (−2, 0)
Point B: Right (positive) one and up zero = (1, 0)
Point C: Left/right zero and up (positive) two = (0, 2)
Point D: Left/right zero and down (negative) three = (0, −3)

Fill in the ordered pair for each point.

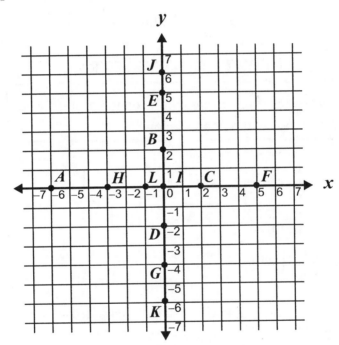

1. point A = (,   )

2. point B = (,   )

3. point C = (,   )

4. point D = (,   )

5. point E = (,   )

6. point F = (,   )

7. point G = (,   )

8. point H = (,   )

9. point I = (,   )

10. point J = (,   )

11. point K = (,   )

12. point L = (,   )

SYMMETRY

Many geometric figures are symmetrical or have **symmetry**. Geometric figures can have three types of symmetry: **reflectional**, **rotational**, and **translational**.

REFLECTIONAL SYMMETRY

A figure has **reflectional symmetry** if you can draw a line through the figure that divides it into two mirror images. The mirror image line is called the **line of symmetry**. Look at the figures below.

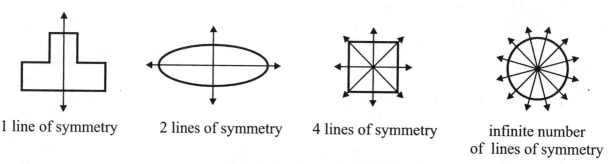

| 1 line of symmetry | 2 lines of symmetry | 4 lines of symmetry | infinite number of lines of symmetry |

ROTATION SYMMETRY

A figure has **rotational symmetry** if the image will lie on top of itself when rotated through some angle other than 360°. Look at the figures below.

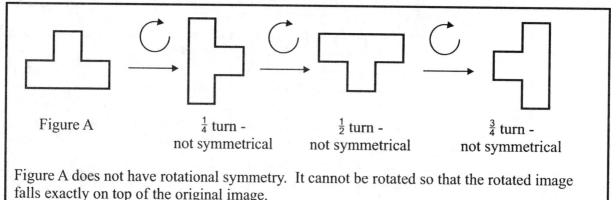

Figure A $\frac{1}{4}$ turn - $\frac{1}{2}$ turn - $\frac{3}{4}$ turn -
                  not symmetrical   not symmetrical   not symmetrical

Figure A does not have rotational symmetry. It cannot be rotated so that the rotated image falls exactly on top of the original image.

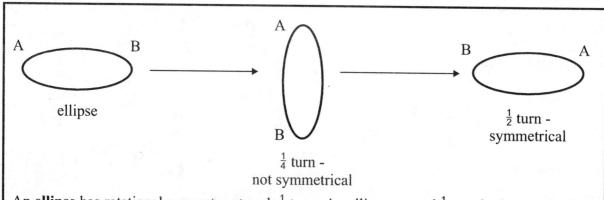

An **ellipse** has rotational symmetry at each $\frac{1}{2}$ turn. An ellipse rotated $\frac{1}{2}$ turn looks exactly like the original ellipse. All the points would lie exactly on top of each other.

200

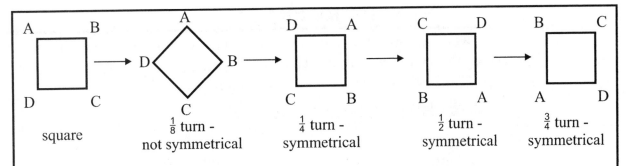

A **square** has rotational symmetry at each $\frac{1}{4}$ turn. A square rotated $\frac{1}{4}$, $\frac{1}{2}$, or $\frac{3}{4}$ looks identical to the original image.

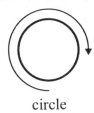

circle

A **circle** has complete rotational symmetry. No matter how much you rotate a circle, the rotated image will always look identical to the original circle.

TRANSLATIONAL SYMMETRY

A geometric pattern has **translational symmetry** if an image can be slid a fixed distance in opposite directions to obtain the same pattern.

This pattern has translational symmetry because when you slide it **horizontally**, it matches the same pattern.

This pattern has translational symmetry because when you slide it **vertically**, it matches the same pattern.

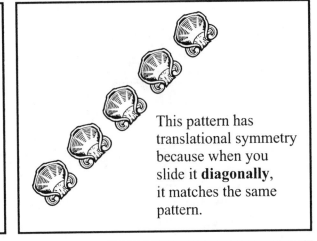

This pattern has translational symmetry because when you slide it **diagonally**, it matches the same pattern.

This pattern **does not** have translational symmetry. The image does not match the same pattern as you slide it in any direction.

SYMMETRY PRACTICE

Match each figure below to the letter that describes its symmetry. Some have more than one answer. Choose all the letters that apply.

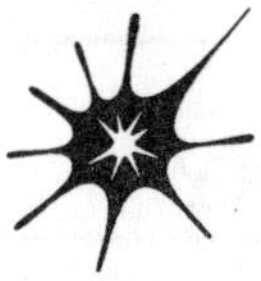

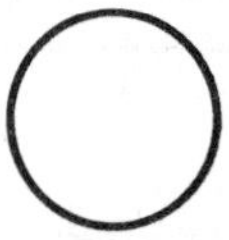

1. _____   A. $\frac{1}{4}$ turn rotational symmetry

2. _____   B. $\frac{1}{2}$ turn rotational symmetry

3. _____   C.   complete rotational symmetry

4. _____   D.   reflectional symmetry

5. _____   E.   translational symmetry

6. _____   F.   not symmetrical

7. _____

8. _____

9. _____

10. _____

11. How many lines of symmetry can be drawn though a regular pentagon?

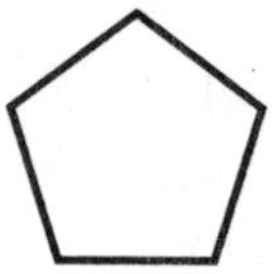

_____

12. How many lines of symmetry can be drawn through the following parallelogram?

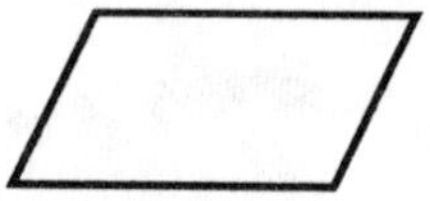

_____

Name the following figures.

1.

_____

2.
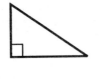

_____

Use the diagram below to answer questions 3 and 4.

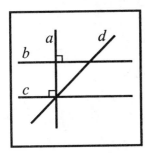

3. Name two lines that are parallel.

_____

4. Name a pair of perpendicular lines.

_____

5. Calculate the perimeter of the following figure.

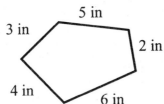

5 in
3 in
2 in
4 in
6 in

_____

Calculate the perimeter and area of the following figures.

6.
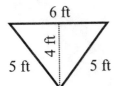

6 ft
4 ft
5 ft   5 ft

$P =$ _____

$A =$ _____

7.
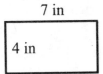

7 in
4 in

$P =$ _____
$A =$ _____

Calculate the circumference and the area of the following circles.

8.
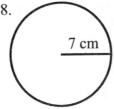

7 cm

Use $\pi = \dfrac{22}{7}$.

$C =$ _____

$A =$ _____

9.
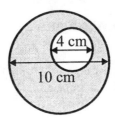

2 ft

Use $\pi = 3.14$.

$C =$ _____

$A =$ _____

10. Use $\pi = 3.14$ to find the area of the shaded part. Round your answer to the nearest whole number.

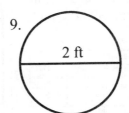

4 cm
10 cm

$A =$ _____

11. Use a ruler to measure the dimensions of the following figure in inches. Find the perimeter.

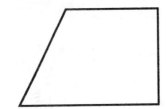

$P =$ _____

12. Use a ruler to measure the dimension of the following figure in centimeters. Find the perimeter and area.

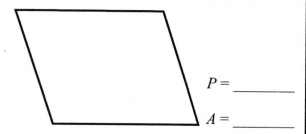

P = _____

A = _____

13. The shaded area below represents Grimes National Park. On the grid below, each square below represents 10 square miles. Estimate the area of Grimes National Park.

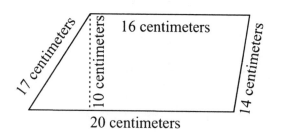

Area is about _____

Find the area of the following figures.

14.

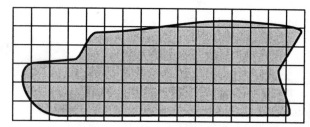

10 inches

6 inches   7 inches

A = _____

15.

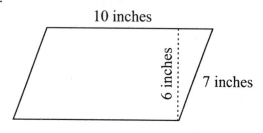

17 centimeters   10 centimeters   16 centimeters   14 centimeters

20 centimeters

A = _____

16.

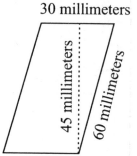

18 inches

12 inches   9 inches   10 inches

10 inches

A = _____

17.

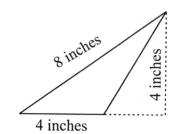

30 millimeters

45 millimeters   60 millimeters

A = _____

18.

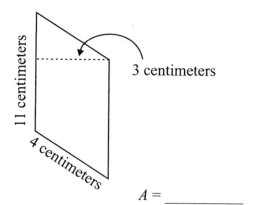

8 inches   4 inches

4 inches

A = _____

19.

11 centimeters

3 centimeters

4 centimeters

A = _____

20. What is the area of a square which measures 8 inches on each side.

A = _____

Record the coordinates and quadrants of the following points.

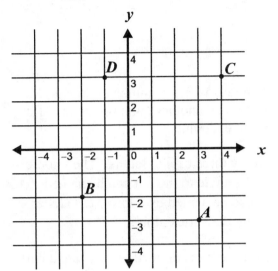

Coordinates   **Quadrant**

21. $A =$ _____ _____

22. $B =$ _____ _____

23. $C =$ _____ _____

24. $D =$ _____ _____

On the same plane above, label these additional coordinates.

25. $E = (0, -3)$

26. $F = (-3, -1)$

27. $G = (4, 0)$

28. $H = (2, 2)$

Answer the following questions.

29. In which quadrant does the point (2, 3) lie? _____

30. In which quadrant does the point (−5, −2) lie? _____

Record the kind(s) of symmetry, if any, shown in each example below.

31. ✿✿✿✿ _____

_____

32. _____

_____

33. ⊕ _____

_____

34. ◎ _____

_____

35. What kind(s) of symmetry does the following figure have? Choose the best answer.

A. reflectional symmetry
B. rotational symmetry
C. reflectional and rotational symmetry
D. no symmetry

36. What kind(s) of symmetry does the following figure have? Choose the best answer.

A. $\frac{1}{4}$ rotational symmetry
B. $\frac{1}{2}$ rotational symmetry
C. complete rotational symmetry
D. no symmetry

Chapter 14

ANGLES

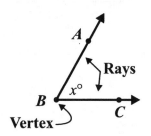

Angles are made up of two rays with a common endpoint. Rays are named by the endpoint B and another point on the ray. Ray $\overrightarrow{BA}$ and ray $\overrightarrow{BC}$ share a common endpoint.

Angles are usually named by three capital letters. The middle letter names the vertex. The angle to the left can be named $\angle ABC$ or $\angle CBA$. An angle can also be named by a lower case letter between the sides, $\angle x$, or by the vertex alone, $\angle B$.

A protractor, , is used to measure angles. The protractor is divided evenly into a half circle of 180 degrees (180°). When the middle of the bottom of the protractor is placed on the vertex, and one of the rays of the angle is lined up with 0°, the other ray of the angle crosses the protractor at the measure of the angle. The angle below has the ray pointing left lined up with 0° (the outside numbers), and the other ray of the angle crosses the protractor at 55°. The angle measures 55°.

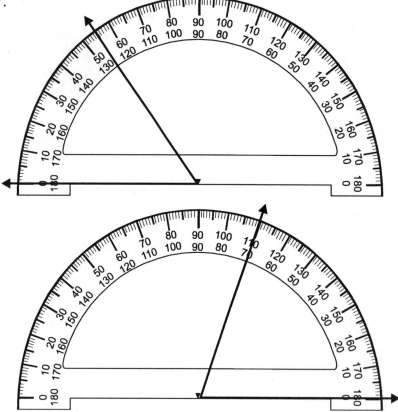

The angle above has the ray pointing right lined up with 0° using the inside numbers. The other ray crosses the protractor and measures the angle at 70°.

TYPES OF ANGLES

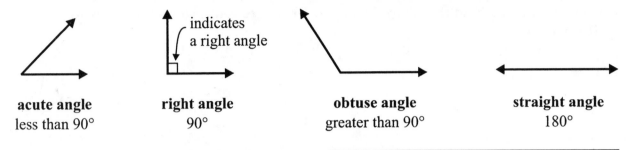

acute angle
less than 90°

right angle
90°

obtuse angle
greater than 90°

straight angle
180°

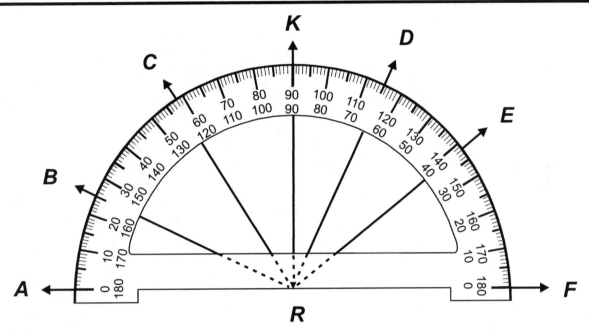

Using the protractor above, find the measure of the following angles. Then, tell what type of angle it is: acute, right, obtuse, or straight.

		Measure	Type of Angle
1.	What is the measure of angle *ARF*?	_____	_____
2.	What is the measure of angle *CRF*?	_____	_____
3.	What is the measure of angle *BRF*?	_____	_____
4.	What is the measure of angle *ERF*?	_____	_____
5.	What is the measure of angle *ARB*?	_____	_____
6.	What is the measure of angle *KRA*?	_____	_____
7.	What is the measure of angle *CRA*?	_____	_____
8.	What is the measure of angle *DRF*?	_____	_____
9.	What is the measure of angle *ARD*?	_____	_____
10.	What is the measure of angle *FRK*?	_____	_____

MEASURING ANGLES

Estimate the measure of the following angles. Then, use your protractor to record the actual measure.

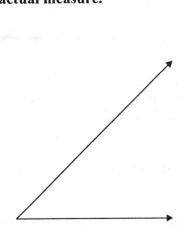

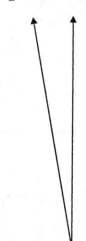

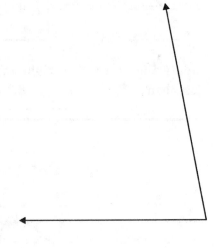

1. Estimate = _____°
    Measure = _____°

4. Estimate = _____°
    Measure = _____°

7. Estimate = _____°
    Measure = _____°

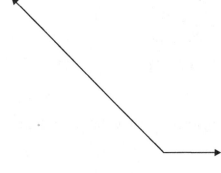

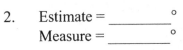

2. Estimate = _____°
    Measure = _____°

5. Estimate = _____°
    Measure = _____°

8. Estimate = _____°
    Measure = _____°

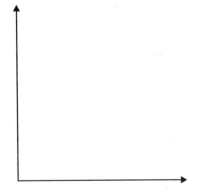

3. Estimate = _____°
    Measure = _____°

6. Estimate = _____°
    Measure = _____°

9. Estimate = _____°
    Measure = _____°

ADJACENT ANGLES

Adjacent angles are two angles that have the same vertex and share one ray. They do not share space inside the angles.

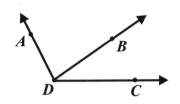

In this diagram, ∠*ADB* is **adjacent** to ∠*BDC*.

However, ∠*ADB* is **not adjacent** to ∠*ADC* because adjacent angles do not share any space inside the angle.

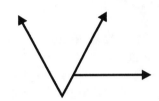

These two angles are **not adjacent.** They share a common ray but do not share the same vertex.

For each diagram below, name the angle that is adjacent to it.

1.

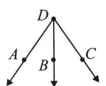

∠*CDB* is adjacent to ∠ _____

2.

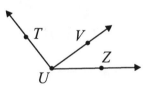

∠*TUV* is adjacent to ∠ _____

3.

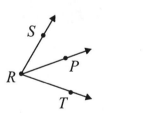

∠*SRP* is adjacent to ∠ _____

4.

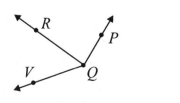

∠*PQR* is adjacent to ∠ _____

5.

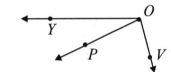

∠*YOP* is adjacent to ∠ _____

6.

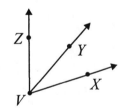

∠*XVY* is adjacent to ∠ _____

7.

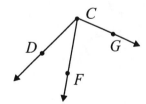

∠*DCF* is adjacent to ∠ _____

8.

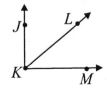

∠*JKL* is adjacent to ∠ _____

VERTICAL ANGLES

When two lines intersect, two pairs of vertical angles are formed. Vertical angles are not adjacent. Vertical angles have the same measure.

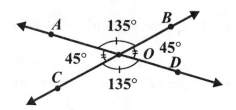

∠AOB and ∠COD are vertical angles. ∠AOC and ∠BOD are vertical angles. **Vertical angles are congruent**. Congruent means they have the same measure.

In the diagrams below, name the second angle in each pair of vertical angles.

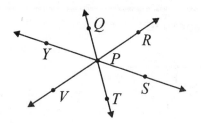
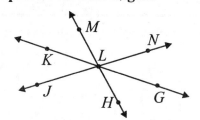

1. ∠YPV _____   4. ∠VPT _____   7. ∠MLN _____   10. ∠GLM _____

2. ∠QPR _____   5. ∠RPT _____   8. ∠KLH _____   11. ∠KLM _____

3. ∠SPT _____   6. ∠VPS _____   9. ∠GLN _____   12. ∠HLG _____

Use the information given to find the measure of each unknown vertical angle.

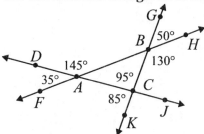
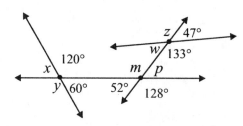

13. ∠CAF = _____   19. ∠x = _____

14. ∠ABC = _____   20. ∠y = _____

15. ∠KCJ = _____   21. ∠z = _____

16. ∠ABG = _____   22. ∠w = _____

17. ∠BCJ = _____   23. ∠m = _____

18. ∠CAB = _____   24. ∠p = _____

COMPLEMENTARY AND SUPPLEMENTARY ANGl

Two angles are **complementary** if the sum of the measures of the angles is 90°.
Two angles are **supplementary** if the sum of the measures of the angles is 180°.

The angles may be adjacent but do not need to be.

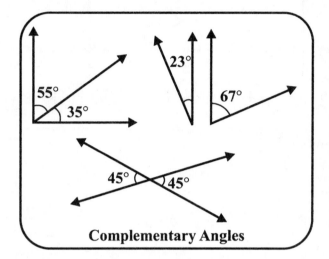

Complementary Angles

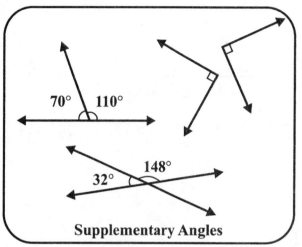

Supplementary Angles

Calculate the measure of each unknown angle.

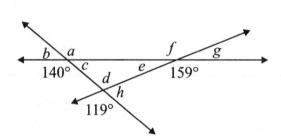

1. ∠a = _____

2. ∠b = _____

3. ∠c = _____

4. ∠d = _____

5. ∠e = _____

6. ∠f = _____

7. ∠g = _____

8. ∠h = _____

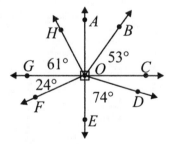

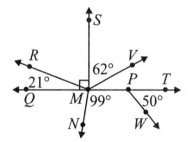

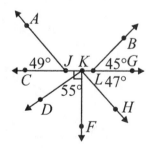

9. ∠AOB = _____

10. ∠COD = _____

11. ∠EOF = _____

12. ∠AOH = _____

13. ∠RMS = _____

14. ∠VMT = _____

15. ∠QMN = _____

16. ∠WPQ = _____

17. ∠AJK = _____

18. ∠CKD = _____

19. ∠FKH = _____

20. ∠BLC = _____

INTERIOR ANGLES OF A TRIANGLE

The three interior angles of a triangle always add up to be 180°.

EXAMPLE 1:

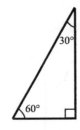 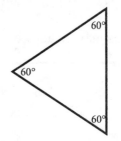

$45° + 45° + 90° = 180°$ $30° + 60° + 90° = 180°$ $60° + 60° + 60° = 180°$

EXAMPLE 2:   Find the missing angle in the triangle.

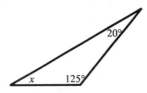

Solution:
$$20° + 125° + x = 180°$$
$$\underline{-20° \ -125° \qquad -20° \ -125°}$$
$$x = 180° - 20° - 125°$$
$$x = 35°.$$

Subtract 20° and 125° from both sides to get x by itself.

The missing angle is 35°.

Find the missing angles in the triangles.

1.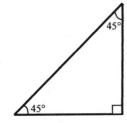

2.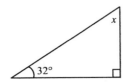

3.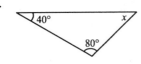

4.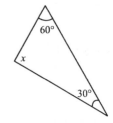

5.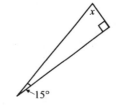

6.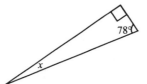

7.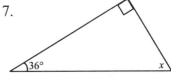

8.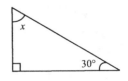

9.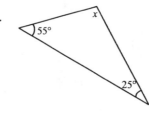

CHAPTER 14 REVIEW

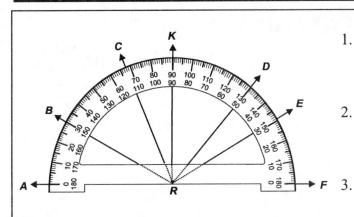

1. What is the measure of ∠DRA?

_____

2. What is the measure of ∠CRF?

_____

3. What is the measure of ∠ARB?

_____

4. Which angle is a supplementary angle to ∠EDF?

_____

5. What is the measure of angle ∠GDF?

_____

6. Which 2 angles are right angles?

_____ and _____

7. What is the measure of ∠EDF?

_____

8. Which angle is adjacent to ∠BAD?

_____

9. Which angle is a complementary angle to ∠HAD?

_____

10. What is the measure of ∠HAB?

_____

11. What is the measure of ∠CAD?

_____

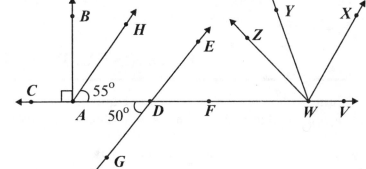

12. What kind of angle is ∠FDA?

_____

13. What kind of angle is ∠GDA?

_____

14. Which angles are adjacent to ∠EDA?

_____ and _____

15. Measure ∠VWX with a protractor.

_____

16. Measure ∠FWY with a protractor.

_____

17. Measure ∠VWY with a protractor.

_____

SOLID GEOMETRY

In this chapter, you will learn about the following three-dimensional shapes.

SOLIDS

 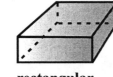  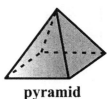

cube   rectangular prism   cone   cylinder   sphere   pyramid

UNDERSTANDING VOLUME

Volume - Measurement of volume is expressed in cubic units such as in^3, ft^3, m^3, cm^3, or mm^3. The volume of a solid is the number of cubic units that can be contained in the solid.

First, let's look at rectangular solids.

EXAMPLE:

How many 1 cubic centimeter cubes will it take to fill up the figure below?

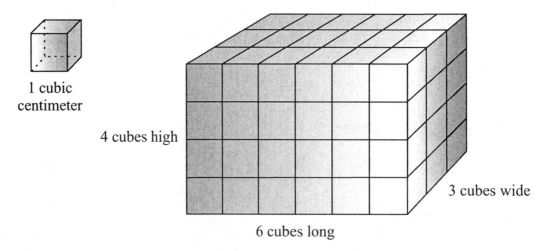

1 cubic centimeter

4 cubes high

6 cubes long

3 cubes wide

To find the volume, you need to multiply the length times the width times the height.

Volume of a rectangular solid = length $\times$ width $\times$ height ($V = l\,w\,h$)

$V = 6 \times 3 \times 4 = 72\ cm^3$

CUBES

EXAMPLE: Find the volume of the cube pictured at the right.

Each side(s) of a cube has the same measure.

The formula for the volume of a cube is: $V = s^3 \ (s \times s \times s)$

Step 1:   Insert measurements from the figure into the formula:

Step 2:   Multiply to solve. $5 \times 5 \times 5 = 125 \text{ cm}^3$

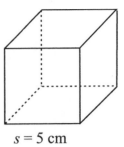

$s = 5$ cm

Note: **Volume is always expressed in cubic units such as in^3, ft^3, m^3, cm^3, or mm^3.**

Answer each of the following questions about cubes.

1.   If a cube is 3 centimeters on each edge, what is the volume of the cube?

   _____

2.   If the measure of the edge is doubled to 6 centimeters on each edge, what is the volume of the cube?

   _____

3.   What if the edge of a 3 centimeters cube is tripled to become 9 centimeters on each edge? What will the volume be?

   _____

4.   How many cubes with edges measuring 3 centimeters would you need to stack together to make a solid 12 centimeter cube?

   _____

5.   What is the volume of a 2 centimeter cube?

   _____

6.   Jerry built a 2 inch cube to hold his marble collection. He wants to build a cube with a volume 8 times larger. How much will each edge measure?

Find the volume of the following cubes.

7.

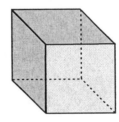

$s = 7$ in $V = $ _____

8.

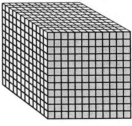

4 ft

4 ft

4 ft $V = $ _____

9.   12 inches = 1 foot

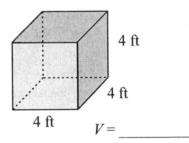

$s = 1$ foot

How many cubic inches are in a cubic foot?   _____

Find the volume of the following rectangular prisms (boxes) and cubes.

1.

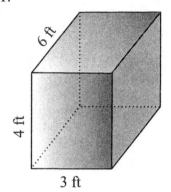

6 ft

4 ft

3 ft

$V =$ _____

4.

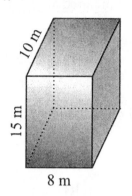

10 m

15 m

8 m

$V =$ _____

7.

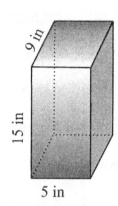

9 in

15 in

5 in

$V =$ _____

2.

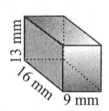

13 mm

16 mm

9 mm

$V =$ _____

5.

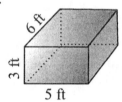

6 ft

3 ft

5 ft

$V =$ _____

8.

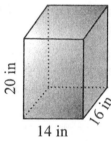

20 in

14 in

16 in

$V =$ _____

3.

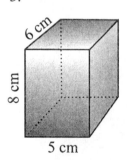

6 cm

8 cm

5 cm

$V =$ _____

6.

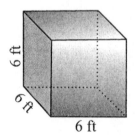

6 ft

6 ft

6 ft

$V =$ _____

9.

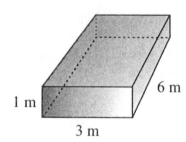

6 m

1 m

3 m

$V =$ _____

VOLUME OF SPHERES, CONES, CYLINDERS, AND PYRAMIDS

To find the volume of a solid, substitute the measurements given for the solid in the correct formula and solve. Volumes are expressed in cubic units such as in^3, ft^3, m^3, cm^3, or mm^3.

Sphere	**Cone**	**Cylinder**
$V = \frac{4}{3}\pi r^3$	$V = \frac{1}{3}\pi r^2 h$	$V = \pi r^2 h$

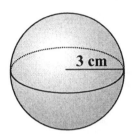 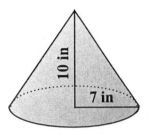 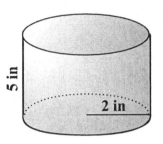

$V = \frac{4}{3}\pi r^3$ $\pi = 3.14$	$V = \frac{1}{3}\pi r^2 h$ $\pi = 3.14$	$V = \pi r^2 h$ $\pi = \frac{22}{7}$
$V = \frac{4}{3} \times 3.14 \times 27$	$V = \frac{1}{3} \times 3.14 \times 49 \times 10$	$V = \frac{22}{7} \times 4 \times 5$
$V = 113.04\ cm^3$	$V = 512.87\ in^3$	$V = 62\frac{6}{7}\ in^3$

Pyramids

$V = \frac{1}{3}Bh$ B = area of rectangular base $V = \frac{1}{3}Bh$ B = area of triangular base

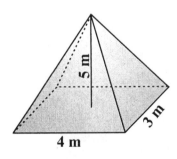   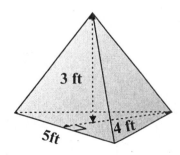

$V = \frac{1}{3}Bh$ $B = l \times w$
$V = \frac{1}{3} \times 4 \times 3 \times 5$
$V = 20\ m^3$

$B = \frac{1}{2} \times 5 \times 4 = 10\ ft^2$
$V = \frac{1}{3} \times 10 \times 3$
$V = 10\ ft^3$

217

Find the volume of each of the following shapes. Use π = 3.14.

1. 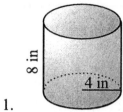 8 in   4 in   V = _____

2. 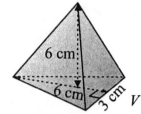 6 cm   6 cm   3 cm   V = _____

3. 5 m   V = _____

4.  8 ft   2 ft   V = _____

5. 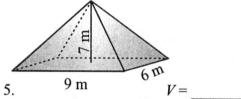 7 m   9 m   6 m   V = _____

6. 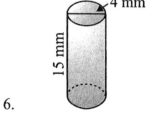 4 mm   15 mm   V = _____

7. 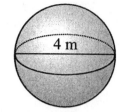 4 m   V = _____

8. 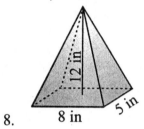 12 in   8 in   5 in   V = _____

9. 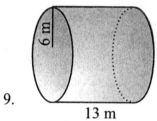 6 m   13 m   V = _____

10. 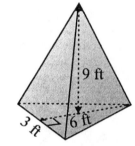 9 ft   3 ft   6 ft   V = _____

11. 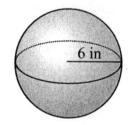 6 in   V = _____

12. 9 in   8 in   V = _____

ESTIMATING VOLUME

Measure the following objects with your ruler and estimate the volume.

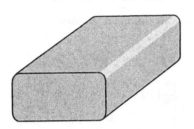

1. Volume is about _____ inches³

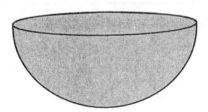

2. Volume is about _____ inches³

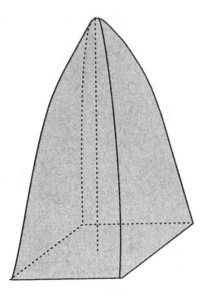

3. Volume is about _____ centimeters³

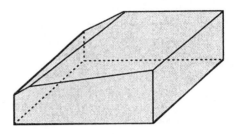

4. Volume is about _____ centimeters³

5. Volume is about _____ centimeters³

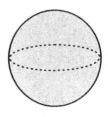

6. Volume is about _____ inches³

SURFACE AREA

The surface area of a solid is the total area of all the sides of a solid.

CUBE

There are six sides on a cube. To find the surface area of a cube, find the area of one side and multiply by 6.

Area of each side of the cube:
$3 \times 3 = 9 \text{ cm}^2$

Total surface area: $9 \times 6 = 54 \text{ cm}^2$

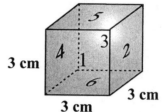

RECTANGULAR PRISM

There are 6 sides on a rectangular prism. To find the surface area, add the areas of the six rectangular sides.

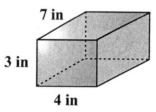

Top and Bottom

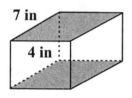

Area of top side:
7 in. $\times$ 4 in. $= 28 \text{ in}^2$
Area of top and bottom:
28 in. $\times$ 2 in. $= 56 \text{ in}^2$

Front and Back

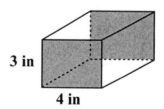

Area of front:
3 in $\times$ 4 in $= 12 \text{ in}^2$
Area of front and back:
12 in $\times$ 2 in $= 24 \text{ in}^2$

Left and Right

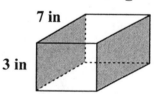

Area of left side:
3 in $\times$ 7 in $= 21 \text{ in}^2$
Area of left and right:
21 in $\times$ 2 in $= 42 \text{ in}^2$

Total surface area: $56 \text{ in}^2 + 24 \text{ in}^2 + 42 \text{ in}^2 = 122 \text{ in}^2$

Find the surface area of the following cubes and prisms.

1.

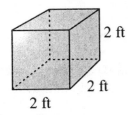

2 ft
2 ft
2 ft

SA = _____

2.

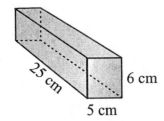

25 cm
6 cm
5 cm

SA = _____

3.

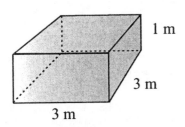

1 m
3 m
3 m

SA = _____

4.

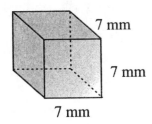

7 mm
7 mm
7 mm

SA = _____

5.

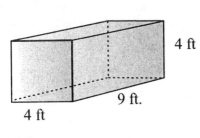

4 ft
9 ft.
4 ft

SA = _____

6.

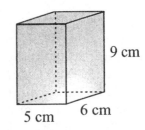

9 cm
5 cm   6 cm

SA = _____

7.

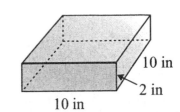

10 in
2 in
10 in

SA = _____

8.

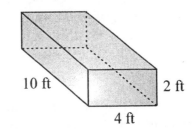

10 ft   4 ft   2 ft

SA = _____

9.

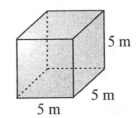

5 m
5 m
5 m

SA = _____

10.

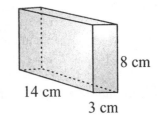

8 cm
14 cm
3 cm

SA = _____

SOLID GEOMETRY WORD PROBLEMS

1. If an Egyptian pyramid has a square base that measures 500 yards by 500 yards, and the pyramid stands 300 yards tall, what would be the volume of the pyramid? Use the formula for volume of a pyramid, $V=\frac{1}{3}Bh$ where B is the area of the base.

   $V = $ _____

2. Robert is using a cylindrical barrel filled with water to flatten the sod in his yard. The circular ends have a radius of 1 foot. The barrel is 3 feet long. How much water will the barrel hold? The formula for volume of a cylinder is $V = \pi r^2 h$. Use $\pi = 3.14$.

   $V = $ _____

3. If a basketball measures 24 centimeters in diameter, what volume of air will it hold? The formula for volume of a sphere is $V = \frac{4}{3}\pi r^3$. Use $\pi = 3.14$.

   $V = $ _____

4. What is the volume of a cone that is 2 inches in diameter and 5 inches tall? The formula for volume of a cone is $V = \frac{1}{3}\pi r^2 h$. Use $\pi = 3.14$.

   $V = $ _____

5. Kelly has a rectangular fish aquarium that measures 24 inches wide, 12 inches deep, and 18 inches tall. What is the maximum amount of water that the aquarium will hold?

   $V = $ _____

6. Jenny has a rectangular box that she wants to cover in decorative contact paper. The box is 10 cm long, 5 cm wide, and 5 cm high. How much paper will she need to cover all 6 sides?

   $SA = $ _____

7. Gasco needs to construct a cylindrical, steel gas tank that measures 6 feet in diameter and is 8 feet long. How many square feet of steel will be needed to construct the tank? Use the following formulas as needed: $A = l \times w$, $A = \pi r^2$, $C = 2\pi r$. Use $\pi = 3.14$.

   $SA = $ _____

8. Craig wants to build a miniature replica of San Francisco's Transamerica Pyramid out of glass. His replica will have a square base that measures 6 cm by 6 cm. The 4 triangular sides will be 6 cm wide and 60 cm tall. How many square centimeters of glass will he need to build his replica? Use the following formulas as needed: $A = l \times w$ and $A = \frac{1}{2}bh$.

   $SA = $ _____

9. Jeff built a wooden, cubic toy box for his son. Each side of the box measures 2 feet. How many square feet of wood did he use to build the toy box? How many cubic feet of toys will the box hold?

   $SA = $ _____

   $V = $ _____

NETS OF SOLID OBJECTS

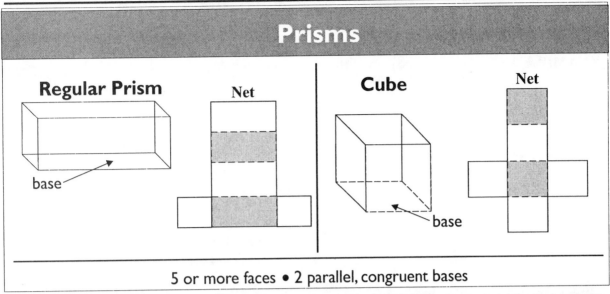

Prisms

Regular Prism

base

Net

Cube

Net

base

5 or more faces • 2 parallel, congruent bases

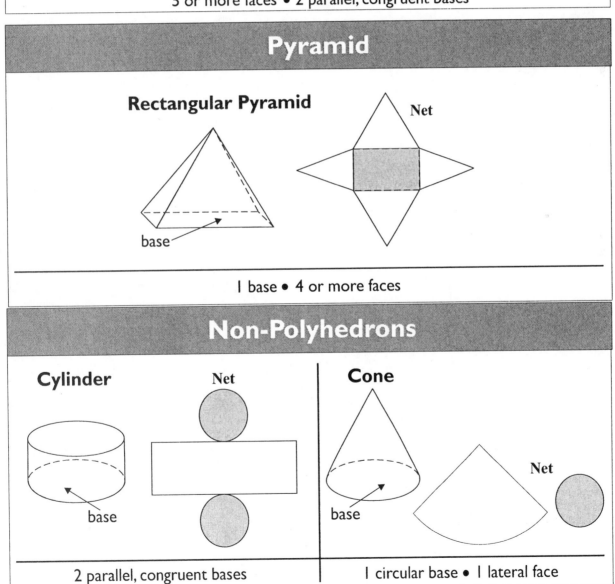

Pyramid

Rectangular Pyramid

Net

base

1 base • 4 or more faces

Non-Polyhedrons

Cylinder

Net

base

2 parallel, congruent bases

Cone

Net

base

1 circular base • 1 lateral face

USING NETS TO FIND SURFACE AREA

A **net** is a two dimensional representation of a three dimensional object. Nets clearly illustrate the plane figures that make up a solid.

EXAMPLE 1: Find the surface area of the figure shown below.

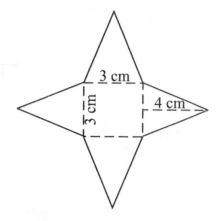

Step 1: Find the area of the 4 triangles.
$A = \frac{1}{2} bh$
$A = \frac{1}{2} \times 3 \times 4 = 6$ cm^2
Area of all 4 triangles = $4 \times 6 = 24$ cm^2

Step 2: Find the area of the base.
$A = lw$
$A = 3 \times 3 = 9$ cm^2

Step 3: Find the sum of the areas of all the plane figures.
Surface Area = 24 cm^2 + 9 cm^2
$SA = 33$ cm^2

EXAMPLE 2: A net for a cone is shown below. Find the surface area.

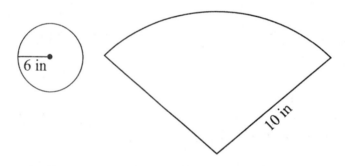

Step 1: Find the area of the base.
$A = \pi r^2$
$A = 3.14 \times 6^2 = 3.14 \times 36 = 113.04$ in^2

Step 2: Find the area of the cone section.
$A = \pi rl$
$A = 3.14 \times 6 \times 10$
$A = 188.40$ in^2

Step 3: Find the sum of the areas of the base and the cone section.
Surface area = 113.04 in^2 + 188.40 in^2
$SA = 301.44$ in^2

**The nets for various solids are given. Find the surface area of the objects.
If needed, use π = 3.14.**

1.

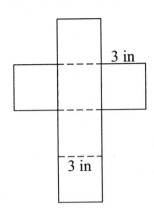

SA = _____

3.

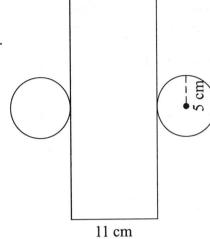

SA = _____

2.

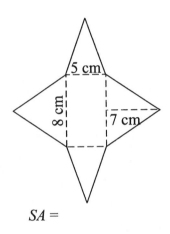

SA = _____

4.

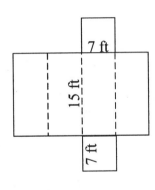

SA = _____

**Using a ruler, measure the dimensions of the following nets to the nearest tenth of a
centimeter, and calculate the surface area of the object. If needed, use π = 3.14.**

5.

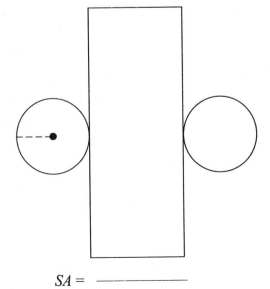

SA = _____

6.

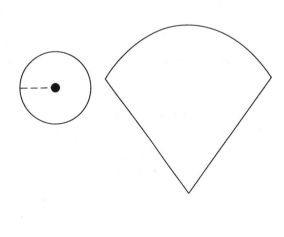

SA = _____

FRONT, TOP, SIDE, AND CORNER VIEWS OF SOLID OBJECTS

Solid objects are 3-dimensional and therefore, are able to be viewed from several perspectives. You should be able to recognize the corner view of a solid given the front, top, and side views. Likewise, you should be able to draw and/or recognize the front, top, and side views of an object given its corner view.

EXAMPLE 1: Draw the front, top, and side views of the object shown below.

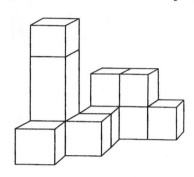

Solution:

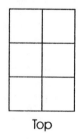

Top

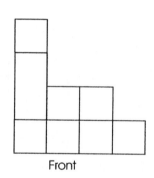

Front

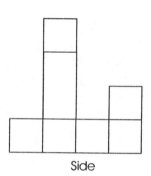

Side

EXAMPLE 2: The top, front, and side views of an object are shown below. How many cubes would it take to build this object.

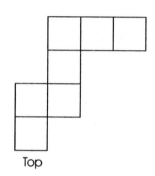

Top

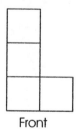

Front

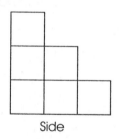

Side

Solution: Draw the object first, and then count the number of cubes used to create the structure.

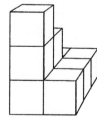

9 cubes are needed to build this object.

226

CHAPTER 15 REVIEW

Find the volume and/or the surface area of the following solids.

1.

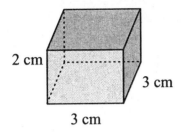

2 cm
3 cm
3 cm

V = _____

SA = _____

2.

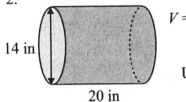

14 in
20 in

$V = \pi r^2 h$

Use $\pi = \dfrac{22}{7}$.

V = _____

3.

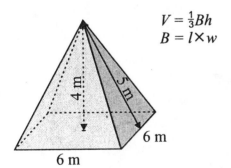

4 m   5 m
6 m
6 m

$V = \frac{1}{3}Bh$
$B = l \times w$

V = _____

4.

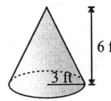

6 ft
3 ft

$V = \frac{1}{3}\pi r^2 h$

Use $\pi = 3.14$.

V = _____

5.

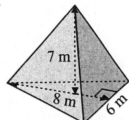

7 m
8 m
6 m

$V = \frac{1}{3}Bh$
B = area of the triangular base

V = _____

6.

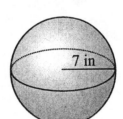

7 in

$V = \frac{4}{3}\pi r^3$

Use $\pi = \dfrac{22}{7}$.

V = _____

7. Use a ruler to estimate the volume of the following figure in cubic centimeters.

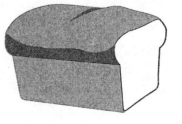

V = _____

8. The sandbox at the local elementary school is 60 inches wide and 100 inches long. The sand in the box is 6 inches deep. How many cubic inches of sand are in the sandbox?

9. A grain silo is in the shape of a cylinder. If the silo has an inside diameter of 10 feet and a height of 35 feet, what is the maximum volume inside the silo? Use $\pi = \dfrac{22}{7}$.

10. A closed cardboard box is 30 centimeters long, 10 centimeters wide, and 20 centimeters high. What is the total surface area of the box?

11. Siena wants to build a wooden toy box with a lid. The dimensions of the toy box are 3 feet long, 4 feet wide, and 2 feet tall. How many square feet of wood will she need to construct the box?

12. The cylinder below has a volume of 240 cubic inches. The cone below has the same radius and the same height as the cylinder. What is the volume of the cone?

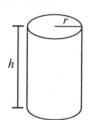 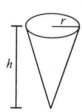

13. Find the volume of the figure below. Each side of each cube measures 4 feet.

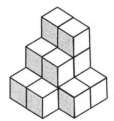

14. The figure below is a net of a rectangular prism. Measure its dimensions to the nearest tenth of a centimeter, and calculate its surface area.

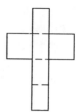

15. Consider the solid shown in the first diagram below. From the following three diagrams, identify and label the front, top, and side views.

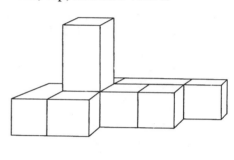

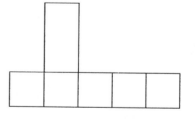

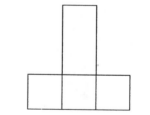

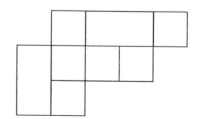

MATHEMATICS PRACTICE TESTS

General Directions

1. Read all directions carefully. When you take the **Minnesota Basic Skills Test,** the directions will be read to you. You should follow along as the directions are read.

2. The questions on this test are followed by several suggested answers. For each question, find the one answer that you think is the best. Then choose that answer on your answer sheet.

3. Choose only one answer for each question. If you change an answer, be sure to erase the first answer completely.

4. **The Minnesota Basic Skills Test is not a timed test. You will be given all the time you need.**

5. The **first 15 questions** are estimation questions and are to be answered **without** the use of a calculator. On the **Minnesota Basic Skills Test,** you will use a sticker to seal questions 1–15 after you complete those questions. Once you have completed questions 1–15, check your answers, and continue with the rest of the test. Do not go back to these questions after you begin with questions 16–76.

6. Calculators are not necessary but may be used for questions 16 through the end of the test.

FORMULA SHEET

Formulas that you may need to work questions on this test are found below.

Area of a square = s^2

Area of a rectangle = lw

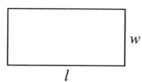

Area of a triangle = $\frac{1}{2} bh$

Area of a circle = πr^2

Circumference = πd or $2\pi r$

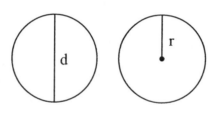

π = Pi = 3.14 or $\frac{22}{7}$

Volume of a cube = $l \times w \times h$

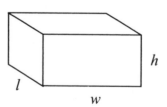

Area of a trapezoid = $\dfrac{(b_1 + b_2) \times h}{2}$

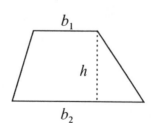

230

PRACTICE TEST 1

1. What are the next two numbers in the numerical pattern below?

    1, 3, 9, 27, 81, ☐, ☐

    A. 115, 162
    B. 135, 189
    C. 180, 372
    D. 243, 729

2. In a survey of 125 randomly selected voters, 80 said they would vote for Steven Gillmor. If 20,000 people in the district vote in the election, approximately how many would be expected to vote for Steven Gillmor?

    E. 3,000
    F. 13,000
    G. 15,000
    H. 16,000

3. In Betty's class there are 16 girls and 14 boys. Which of these is the correct ratio of girls to the total number of students in the class?

    A. 14 to 16
    B. 16 to 14
    C. 14 to 30
    D. 16 to 30

4. Which of these will provide the best estimate of the area of this circle?

    E. $3^2 \times (10 \div 2)$
    F. $3 \times (10 \div 2)^2$
    G. 3×10^2
    H. $(3 \times 10)^2$

    9.7 centimeters

5. Joan has four packages to mail. Their individual weights are 4.11 pounds, 9.87 pounds, 7.61 pounds, and 5.37 pounds. Which is the best estimate of the total weight of the four packages?

    A. 25 pounds
    B. 27 pounds
    C. 28 pounds
    D. 29 pounds

6. As shown in the table, the monthly rent of an apartment depends on the number of bedrooms.

MONTHLY RENT

Bedrooms	Rent
1	$550
2	$625
3	$700

If the pattern is extended, which of these is the likely cost of a 4-bedroom apartment?

E. $725
F. $750
G. $775
H. $800

7. Kevin counted the number of computers in 53 classrooms. The bar graph shows the results of his count.

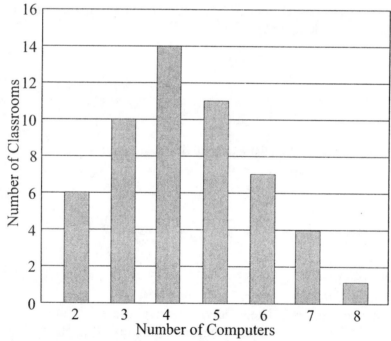

How many classrooms had less than 5 computers?

A. 14
B. 15
C. 20
D. 30

8.   Don's paycheck was $271.93 before taxes. If he gets paid $6 per hour, estimate how many hours he worked.

        E.   35 hours
        F.   40 hours
        G.   45 hours
        H.   50 hours

9.   What is the approximate perimeter of the rectangle below?

6.3 meters

4.6 meters

        A.   20 meters
        B.   22 meters
        C.   28 meters
        D.   30 meters

10.   If you buy 4 cases of soda for a party at $3.10 each, approximately how much will you spend before taxes?

        E. $12.00
        F. $13.00
        G. $14.00
        H. $16.00

11.   Steven is moving and needs to rent a moving truck to get his belongings from his old apartment to his new house. The truck costs $29.95 to rent for 24 hours plus $0.65 per mile that he drives. If he guesses that he will drive at most 25 miles the whole time he has the truck, estimate how much he will spend to rent the truck.

        A. $35.00
        B. $45.00
        C. $55.00
        D. $65.00

12.   A pair of pants costs $45 at the regular price, but the pants are on sale for 25% off. How much will be spent on these pair of pants before taxes?

        E. $11.25
        F. $22.50
        G. $35.00
        H. $33.75

13.   At the new diner that opened up in town, it costs $23.00 for two people's dinners. How
       much will be paid for a dinner of four, if tax is 5%?

       A. $24.15
       B. $46.00
       C. $48.30
       D. $96.60

14.   Stephanie made a 76 on her first test, an 86 on her second test, and a 94 on her fourth
       test. If she knows that her average is 84, what was the grade of her third test?

       E.   80
       F.   84
       G.   90
       H.   96

15.   Henry has $521.27 in his checking account. If he writes a check for $75.96, estimate
       how much is in his checking account after the check was written.

       A. $420.00
       B. $430.00
       C. $445.00
       D. $460.00

**If you want to check your answers to Questions 1–15,
you may do so now. After you have checked your
answers, continue with the rest of the test. On the
Minnesota Basic Skills Test, you will seal these pages
to prevent you from going back to them. You may
use a calculator for the rest of the test, but you may
not go back and check your answers using a
calculator on Questions 1–15.**

16. It takes $1\frac{2}{3}$ yards of fabric to cover one chair. If Shung Kim bought 12 yards of fabric, how many yards would she have left if she covered 6 chairs?

    E. 2 yards
    F. $4\frac{1}{3}$ yards
    G. $6\frac{2}{3}$ yards
    H. 10 yards

17. Howard is buying bread to make 64 san~~d~~ ~~requires 2 slices of~~ bread. A loaf of bread contain~ ~~es of bread will~~ Howard need to buy?

    A. 6
    B. 7
    C. 8
    D. 9

18. Aunt Betty sent Thomas $~~spent $16.78 for~~ his mother's gift, $9.90 for ~~for his twin~~ sisters. How much can he s~

    E. $17.66
    F. $35.32
    G. $43.61
    H. $45.61

19. The posted speed limit is 704 inches/second. What is the speed limit in miles per hour? There are 5280 feet in 1 mile.

    A. 30
    B. 40
    C. 55
    D. 70

20. The number 0.000045 can also be represented as:

    E. 4.5×10^{6}
    F. 45×10^{-7}
    G. 4.5×10^{-5}
    H. 45 thousandths

21. Which of these expressions represents the average rate in miles per hour between 7 AM and 11 AM?

Time	Odometer Reading
7 AM	20825
11 AM	20965

A. $\dfrac{20825 - 20965}{11 - 7}$

B. $\dfrac{20825 - 7}{20965 - 11}$

C. $\dfrac{20965 - 11}{20965 - 7}$

D. $\dfrac{20965 - 20825}{11 - 7}$

22. The school counselor asked 60 students, "What is your favorite class?" The results are shown in the circle graph (pie chart).

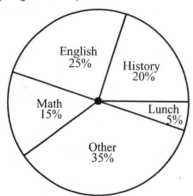

How many students said "Lunch" was their favorite subject?

E. 3
F. 5
G. 8
H. 18

236

23. Florence follows Great Aunt Emma's instructions for making coffee in urns of varying sizes.

Capacity of the Coffee Urn	Number of Scoops of Coffee
4 quarts	18 scoops
6 quarts	26 scoops
10 quarts	42 scoops

Which of these graphs correctly plots the number of scoops as a function of the capacity of the urn?

A.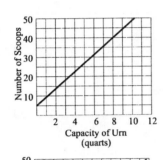

B.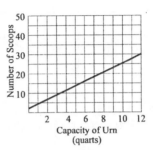

C.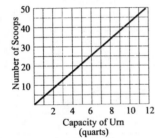

D.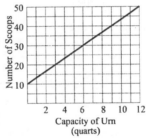

24. The linear relationship between miles and kilometers is shown in the graph below.

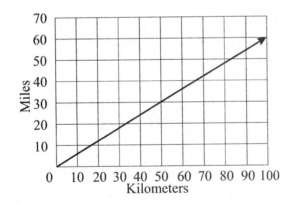

Gregory drove 40 kilometers. About how many miles did he drive?

E.   25 miles
F.   30 miles
G.   55 miles
H.   65 miles

25. Simplify: 9^2

    A.   3
    B.   4
    C.   18
    D.   81

26. Flying at an altitude of 500 meters, Cheryl notices the air temperature outside the plane is 12°C. When the plane climbs to an altitude of 1500 meters, the outside temperature reading is 5°C. Which of these graphs depicts temperature as a linear function of altitude to represent this situation?

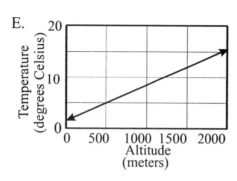

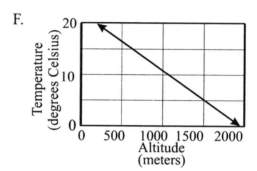

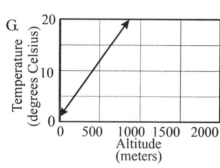

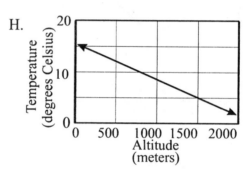

27. There are 18 word-processor keyboards in the library. Each shelf of the storage rack will hold 4 keyboards. How many shelves will be required to hold all eighteen keyboards?

    A.   4
    B.   5
    C.   62
    D.   72

28. The perimeter of a farm is 400 feet. The width of the farm is one-fourth the length of the farm. Find the width and length.

    E. $W = 25, L = 175$
    F. $W = 40, L = 160$
    G. $W = 50, L = 150$
    H. $W = 60, L = 140$

238

29.   As shown in the table, the cost of a repair job depends on the number of hours required for the repair.

COST OF REPAIR

Hours	Cost
1	$72.50
2	$95.00
3	$117.50
4	$140.00

If the repair requires 5 hours, which of these is the cost?

A. $162.50
B. $175.00
C. $177.50
D. $185.00

30.   Janice is comparing the price of three brands of olive oil. Which brand is the best buy?

Olive Oil	Size (milliliters)	Price
Brand X	709 mL	$10.99
Brand Y	500 mL	$8.65
Brand Z	442 mL	$4.99

E.   Brand X is the least expensive per milliliter
F.   Brand Y is the least expensive per milliliter
G.   Brand Z is the least expensive per milliliter
H.   Cannot be determined

31.   The pie chart shows the percentage of time a particular college student spends on various activities during a typical weekday.

What percentage of the student's time is spent in studying and classes?

A.   20%
B.   27%
C.   30%
D.   32%

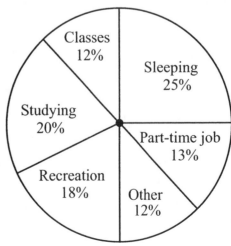

32. To make a disinfecting solution, Alana mixed 2 cups of bleach with 5 cups of water. What is the ratio of bleach to the total amount of disinfecting solution?

E. 2 to 3
F. 2 to 5
G. 2 to 7
H. 2 to 10

33. Which of these is the best estimate of the volume of this box?

A. 550 cubic feet
B. 600 cubic feet
C. 750 cubic feet
D. 1200 cubic feet

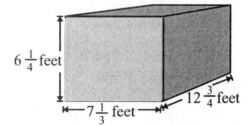

$6\frac{1}{4}$ feet

$7\frac{1}{3}$ feet

$12\frac{3}{4}$ feet

34. Kenneth is planning to build a sailboat that will be 6.5 meters long. In the plans, the length of the sailboat is 130 millimeters. What is the ratio of the length of the actual sailboat to the length of the sailboat in the plans?

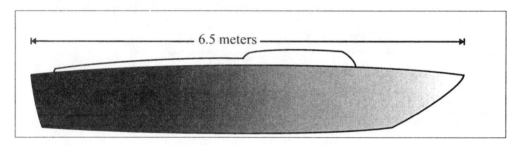

6.5 meters

E. 5 to 1
F. 20 to 1
G. 50 to 1
H. 200 to 1

35. Cindy grew 5 giant pumpkins this year. The weights of the pumpkins, in pounds, are listed below.

57, 91, 77, 67, 98

What is the mean weight of Cindy's pumpkins?

A. 67 pounds
B. 68 pounds
C. 77 pounds
D. 78 pounds

36. The bar graph shows the average daily high temperature at Point Henry each month for the year 2003.

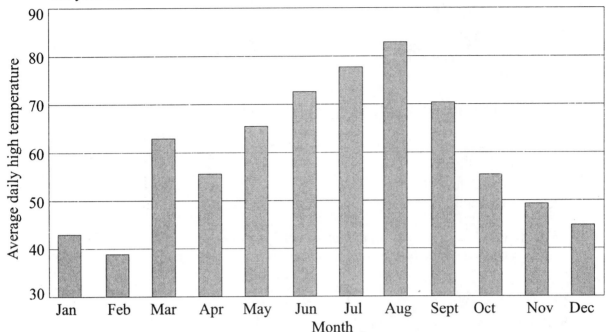

Which of these statements is false?

E.   During 6 of the months the average daily high exceeded 60°.
F.   The average daily high for the entire year 2003 exceeded 80°.
G.   The average daily high for August was higher than for July.
H.   The average daily high for March exceeded 60°.

37. Ms. Madden asked the students in her second period algebra class to count the number of coins each of them had. The results are shown in the bar graph.

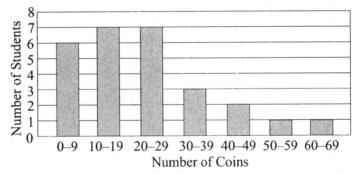

How many students had 20 or more coins?

A.   7
B.   9
C.   12
D.   14

38.   Which of these sets of numbers is ordered from least to greatest?

E. $\frac{1}{4}, -\frac{1}{3}, 0.4, -0.5, 3$
F. $0.4, -\frac{1}{3}, 3, -0.5, \frac{1}{4}$
G. $-0.5, -\frac{1}{3}, \frac{1}{4}, 0.4, 3$
H. $-0.5, -\frac{1}{3}, 0.4, \frac{1}{4}, 3$

39.   In the number 105.728 the digit 8 is in the _____ place.

A.   thousands
B.   ones
C.   hundredths
D.   thousandths

40.   In the figures below, an edge of the larger cube is twice as big as an edge of the smaller cube. What is the ratio of the volume of the smaller cube to that of the larger cube?

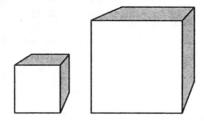

E.   1:2
F.   1:4
G.   1:8
H.   1:16

41.   To find the height of a monument, John measured the length of his shadow and the length of the shadow of the monument. John is 72 inches tall and cast a 40 inch shadow. The monument cast a 24 foot shadow. How tall is the monument? Round to the nearest foot.

 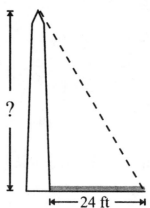

72 in   40 in   ?   24 ft

Note: Figures are not drawn to scale.

A.   13 feet
B.   22 feet
C.   43 feet
D.   84 feet

42. How many lines of symmetry can be drawn through the following figure? Choose the best answer.

    E.   2
    F.   4
    G.   infinite
    H.   none

43. Brent and Andrew are training, preparing to someday swim the English Channel. They swam across Blue Lake from Eagle Point to Clark's Resort. Approximately how far did they swim?

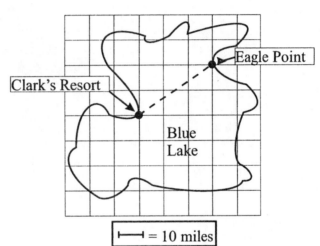

    A.   20 miles
    B.   30 miles
    C.   40 miles
    D.   50 miles

44. Which of these sets of numbers is ordered from least to greatest?

    E. $-\frac{1}{3}, -0.3, \frac{2}{13}, \frac{1}{6}, 0.17$

    F. $-0.3, -\frac{1}{3}, \frac{2}{13}, \frac{1}{6}, 0.17$

    G. $\frac{1}{6}, \frac{2}{13}, 0.17, -\frac{1}{3}, -0.3$

    H. $-0.3, -\frac{1}{3}, 0.17, \frac{2}{13}, \frac{1}{6}$

45.   Justin recorded the weights of 6 wrestlers. Their weights, in kilograms, are given below.

66, 97, 52, 53, 76, 105

What is the median weight of the 6 wrestlers?

A.   52.5 kilograms
B.   71.0 kilograms
C.   85.5 kilograms
D.   86.5 kilograms

46.   To determine the proper size of air conditioning unit for the classroom represented below, Ms. Waters calculated the volume of the classroom.

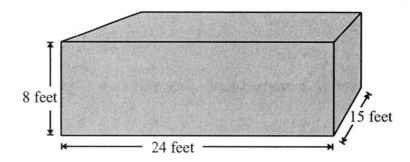

What is the volume of the classroom?

E.   47 cubic feet
F.   672 cubic feet
G.   1334 cubic feet
H.   2880 cubic feet

47.   Justin measured the heights of 10 basketball players. Their heights, in inches, are given below.

71, 82, 72, 78, 73, 76, 72, 75, 73, 78

What is the median height of the 10 basketball players?

A.   73 inches
B.   74 inches
C.   75 inches
D.   76 inches

48. Bobby went to the mall to buy a sweater for his mom for Mother's Day. He found a soft, blue one that she will love for $63. The store was having a Mother's Day sale, so everything was $\frac{1}{3}$ off. How much did the sweater cost before tax?

    E. $21.00
    F. $19.50
    G. $42.00
    H. $43.50

49. Sarah bought groceries yesterday. She spent $4.85 on strawberries, $2.50 on milk, and $1.15 on bread. If sales tax is 6%, how much did Sarah spend altogether?

    A. $9.01
    B. $9.50
    C. $10.01
    D. $10.50

50. Darren works for the Burger Hut by his house. He has worked there for 1 year and 2 months, and his boss decided to promote him. With his new promotion he will be able to save enough money to buy a new car. The car he wants costs $6000. If Darren starts saving now, how many weeks will it be before he can buy the car? What additional information do you need to complete this problem?

    E.   How much he gets paid every week
    F.   How much his raise was
    G.   How much money he already has
    H.   How much he can save each week

51. Byron ate dinner with his wife at their favorite restaurant. The bill was $32.90. If Byron always tips 20%, what is the total that he is going to pay for dinner?

    A. $26.32
    B. $36.19
    C. $38.00
    D. $39.48

52. It is 65°F outside, but tomorrow it is predicted to be 81°F. What is the difference between the temperature today and tomorrow?

    E.   −16°F
    F.   6°F
    G.   16°F
    H.   Cannot be determined

53. A large frozen coffee at the local coffee shop cost $3.90. Tax is always 5%. If you buy two large frozen coffees and give the cashier $20, how much change will you get back?

    A. $8.19
    B. $10.75
    C. $11.81
    D. $12.42

54. Mrs. Brown went to a yard sale and bought 1 item for $14.85, 1 item for $8.00, and 2 items for $5.25 each. She wrote one check for the total amount. If Mrs. Brown had $234.78 in her checking account before going to the yard sale, what will be the balance in her checkbook after her purchase?

E. $33.35
F. $201.43
G. $206.68
H. $268.13

55. Which equation or inequality is false?

A. $\frac{1}{12} > 10\%$
B. $.25 = \frac{1}{4}$
C. $67\% > \frac{5}{8}$
D. $14\% < .15$

56. What would be the appropriate estimate for the length of a standard classroom?

E. 45 yards
F. 55 miles
G. 100 meters
H. 1000 millimeters

57. What is the best estimate for $3624 \div 59$?

A. 60
B. 65
C. 545
D. 600

58. Sam and Kevin swam a tie-breaking race today to determine which of the two would continue onto the state semi-finals. The two have always been fierce competitors with finishing times varying by milliseconds. During the race, the stopwatch of the timer recording Kevin's time failed to function properly. Do you think it would be appropriate to estimate Kevin's time to determine who should continue onto the semi-finals?

E. It would be appropriate if both boys' times were estimated to the nearest second.
F. No, it would not be appropriate because the competition is very close, so every millisecond counts.
G. Yes, it would be appropriate if no one knew that Kevin's time was estimated and Sam's time was actual.
H. No, it is only appropriate to estimate volumes and lengths, not time.

246

59. What is the best estimate for the weight of two twin eight year old girls?

    A.   5 meters
    B.   15 ounces
    C.   110 pounds
    D.   1000 kilograms

60. Estimate: $7.6204 - 3.045$

    E.   4
    F.   5
    G.   6
    H.   7

Use the protractor for questions 61 and 62.

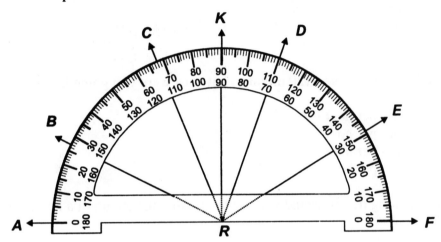

61. What is the measurement of $\angle FRD$?

    A.   40°
    B.   70°
    C.   110°
    D.   180°

62. What is the measurement of $\angle FRE + \angle ERC$?

    E.   70°
    F.   81°
    G.   112°
    H.   220°

63.   What is the length of the line in centimeters?

    A.   8.6 centimeters
    B.   8.7 centimeters
    C.   8.8 centimeters
    D.   9.0 centimeters

64.   What is the area of the rectangle below if the length is 12 yards and the width is 8 yards?

    E.   20 yards^2
    F.   40 yards^2
    G.   80 yards^2
    H.   96 yards^2

65.   If you traveled 56 miles in 120 minutes, how fast were you going in miles per hour?

    A.   28 miles per hour
    B.   47 miles per hour
    C.   56 miles per hour
    D.   84 miles per hour

66.   Which measuring device would you use to measure the length of a kitchen table?

    E.   odometer
    F.   yard stick
    G.   thermometer
    H.   gauge

67.   Which of the following equals 10 grams?

    A.   0.001 milligrams
    B.   0.01 kilograms
    C.   1 milligram
    D.   1000 kilograms

68.   Which of the following is an appropriate measurement for the temperature of pool water?

    E.   5°C
    F.   45°F
    G.   70°F
    H.   100°C

69. There are two red jelly beans, six blue jelly beans, and four green jelly beans. What is the probability that you will pick a green jelly bean on the first try?

   A.　$\frac{1}{12}$
   B.　$\frac{1}{4}$
   C.　$\frac{1}{3}$
   D.　$\frac{1}{2}$

70. If this shape's sides were folded upward at the dotted lines, what three-dimensional object would it make?

   E.　cube
   F.　cylinder
   G.　cone
   H.　pyramid

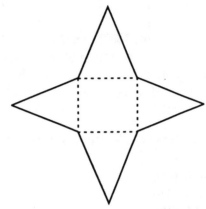

71. What is the name of this shape?

   A.　pentagon
   B.　hexagon
   C.　trapezoid
   D.　parallelogram

72. Consider a cylinder standing upright on its base. The shape of the top of the cylinder is a

   E.　point.
   F.　triangle.
   G.　square.
   H.　circle.

73. Michele and her sister, Karen, are having a disagreement. They have been arguing on whether to go to the zoo or to the aquarium. If they decide to flip a coin to decide, what is the probability that it will land on heads?

   A.　$\frac{1}{2}$
   B.　$\frac{2}{3}$
   C.　1
   D.　cannot be determined

74. A survey was taken of how many people ate school lunch each day. The survey was taken for two weeks. Here are the results:

85, 77, 85, 62, 90, 88, 88, 85, 75, 95

What is the median of this data?

E. 83
F. 85
G. 87
H. 95

75. Henry is the manager of the local electronic store in his hometown. A survey was taken that discovered that 4 out 5 people prefer the 42 inch TV to the 35 inch TV. This week Henry needs to order stock for the store, and he is going to base his purchase on the new survey that was taken. If he is allowed to order 50 TV sets, how many of each kind should he buy?

A. 50 - 42 inch TV's
   0 - 35 inch TV's
B. 10 - 42 inch TV's
   40 - 35 inch TV's
C. 40 - 42 inch TV's
   10 - 35 inch TV's
D. 20 - 42 inch TV's
   30 - 35 inch TV's

76. A Republican politician running for mayor wants to take a poll to determine if he is the favored candidate in the upcoming election. The mayoral candidate is scheduled to give a speech not at all related to the election at an upcoming town celebration. For convenience sake, his staff polls all of the people in attendance at the speech as to who they are planning on voting for in the upcoming mayoral election. The staff reports to the candidate that, based on the recent poll, he is favored by the voters. Are the results of this poll most likely accurate or are they most likely misleading?

E. The results of the survey are accurate as long as at least 100 people were polled.
F. The results of the poll are most likely misleading because the sample of people surveyed were voluntarily attending a speech given by the candidate and do not necessarily represent the population of the town.
G. The results of the poll are most likely accurate because people are generally honest in declaring who they are going to vote for in elections.
H. The results of the poll are misleading because recent advertisements are encouraging people to vote, creating a more opinionated population than usual.

 **This is the end of the test.**

PRACTICE TEST 2

1. Joseph goes to a fast food restaurant and orders a hamburger for $1.79, fries for $1.39, and a large shake for $1.99. What is a reasonable amount to give the cashier?

    A. $ 5.00
    B. $11.00
    C. $ 3.00
    D. $ 6.00

MENU

Beef Pie Surprise	$ 6.50
Pork Sandwich	$ 4.75
Ribs in Sauce	$11.45
Bar-B-Q Chicken	$ 5.50
Beverages:	
Tea, Coffee, or Milk	$ .90
Ice Cream	$ 1.85

2. Mary had chicken, and Tony had a pork sandwich at a nearby restaurant. Each had one beverage. What is the estimated cost of the dinner for the two of them?

    E. $ 8.00
    F. $14.00
    G. $12.00
    H. $11.00

3. Using a ruler, estimate the area of the following figure in square inches.

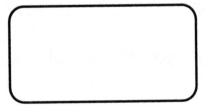

    A.   1 square inch
    B.   2 square inches
    C.   4 square inches
    D.   6 square inches

4. A 10-pound bag of fertilizer will cover 50-square feet of garden space. How many 10 pound bags of fertilizer would you need to cover a garden that is 480 square feet?

    E.   9
    F.   10
    G.   20
    H.   48

5. Phil needed 52 eggs for the Spring Festival egg toss contest. He purchased the eggs at a store that only sold eggs by the dozen. How many dozen eggs did he purchase?

    A.   4
    B.   5
    C.   10
    D.   Not enough information is given.

6. The problem below was done on the calculator, but the decimal point was missing. Estimate to find the correct answer.

$$79.56 \div 4.7$$

    E.   1.6927659
    F.   16.927659
    G.   169.27659
    H.   1692.7659

7. It took Melanie 161 minutes to do her homework last night. About how many hours did she spend on her homework?

    A. $2\frac{2}{3}$ hours
    B.   2 hours
    C. $3\frac{1}{2}$ hours
    D. $1\frac{1}{2}$ hours

8. Emily had $100.00 to spend. She spent $37.85 on clothes. About how much change should she receive from $100.00?

    E. $30.00
    F. $40.00
    G. $60.00
    H. $70.00

9. Use a ruler to estimate the area of the figure below.

    A.   1 square centimeter
    B.   2 square centimeters
    C.   3 square centimeters
    D.   4 square centimeters

10. A bricklayer uses 7 bricks per square foot of surface covered. If he covers an area 28 feet wide and 10 feet high, about how many bricks will he use?

    E.   40
    F.   2,000
    G.   2,800
    H.   3,000

11. The problem below was done on the calculator, but the decimal point was missing. Estimate to find the correct answer.

$$58.59 \div 6.4$$

    A.   9154.6875
    B.   915.46875
    C.   91.546875
    D.   9.1546875

12. According to the phone bill, Pam spent 224 minutes talking to her boyfriend, John, last month. About how many hours was that?

    E. $2\frac{1}{3}$ hours
    F. $2\frac{3}{4}$ hours
    G. $3\frac{3}{4}$ hours
    H.   22 hours

13. Which of the following choices is the best estimate for the width of a standard door?

    A.   9 meters
    B.   9 kilometers
    C.   90 centimeters
    D.   90 millimeters

14. Hanna earns 12% commission on any jewelry sales she makes. About how much is her commission on a $45 sale?

  E. $1.00
  F. $4.00
  G. $5.00
  H. $12.00

This drawing shows the city boundaries of Johnsonville.

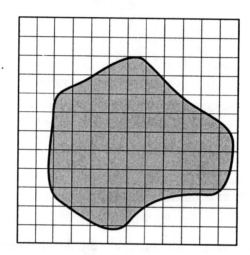

15. Estimate the area inside the city boundaries from the scale diagram above. Each square equals one square mile.

  A. 50 square miles
  B 60 square miles
  C. 70 square miles
  D. 80 square miles

If you want to check your answers to Questions 1–15, you may do so now. After you have checked your answers, continue with the rest of the test. On the Minnesota Basic Skills Test, you will seal these pages to prevent you from going back to them. You may use a calculator for the rest of the test, but you may not go back and check your answers using a calculator on Questions 1–15.

16. Herbie spent a total of $3.75 on popcorn at the movies. What was his change from a $10 dollar bill?

    A.   3 one dollar bills and 3 quarters
    B.   1 five dollar bill, 2 one dollar bills and 1 quarter
    C.   1 five dollar bill, 1 one dollar bill and 1 quarter
    D.   1 five dollar bill and 3 quarters

17. Alphonso saw a stereo on sale for $\frac{1}{3}$ off the regular price of $630.00. How much money could he save if he bought the stereo on sale?

    E. $210.00
    F. $410.00
    G. $600.00
    H. $630.00

18. Which fabric is the cheapest?

    A.   3 yards for $ 2.00
    B.   4 yards for $10.00
    C.   2 yards for $ 4.00
    D.   3 yards for $ 2.50

Name	Average Community Service Hours Per Week	Average Community Service Hours Per Year
Walter	20	1,040
Monisha	15	780
Thomas	10	520
Nyia	8	410

19. Looking at the chart above, how many community service hours will Thomas have in two and one half years?

    E.   520 hours
    F.   1,040 hours
    G.   1,300 hours
    H.   780 hours

```
┌─────────────────────────────────────────┐
│              LIVING ROOM SET              │
│               $1,800.00 cash              │
│                    or                     │
│    Pay $90.00 down and $100.00 a month    │
│              for 24 months                │
└─────────────────────────────────────────┘
```

20. Look at the advertisement above. Mrs. Ortega will be buying the living room set on the payment plan. How much more will she pay by buying on credit?

   A. $ 600.00
   B. $1,200.00
   C. $ 390.00
   D. $ 690.00

21. It is $-3°$ outside right now, and tonight the temperature is expected to drop another $21°$. How cold is it expected to get?

   E. $-18°$
   F. $18°$
   G. $-24°$
   H. $-21°$

22. On a recent test, 80% of the math class got A's. What fraction of the class is that?

   A. $\frac{1}{4}$
   B. $\frac{5}{6}$
   C. $\frac{5}{8}$
   D. $\frac{4}{5}$

23. Elaine sells bread she bakes at home in three local food stores. She figured it is costing her 89¢ per loaf to make the bread, and she sells it for $2.25 per loaf. She works 25 hours a week. To find out how much Elaine makes an hour, you also need to know:

   E.   how many days she works in a week.
   F.   how many days a loaf of bread can stay on the shelf before the expiration date.
   G.   how many loaves she sells in a week.
   H.   how far away the stores are from her home.

24. Jeff was making $6.25 per hour. His boss gave him a $.75 per hour raise. What percent raise did Jeff get?

   A. 12%
   B. 25%
   C. 40%
   D. 70%

25. What speed would a plane average on a trip of 4,000 miles for 6 hours?

    E.   150.0 miles per hour
    F.   666.6 miles per hour
    G.   366.6 miles per hour
    H. 1500.0 miles per hour

26. George started on a trip with the odometer reading 17,853.0 miles. He ended his trip with the odometer reading 18,453.0 miles. He used 40 gallons of gas. How many miles to the gallon did George get?

    A.   15
    B.   16
    C.   122
    D.   14

27. Mr. Mendoza was offered a 15% discount when he purchased a $115,000.00 home. What is the amount of the discount on Mr. Mendoza's home?

    E. $15,000.00
    F. $17,250.00
    G. $16,500.00
    H. $17,520.00

28. Michael borrowed $4,500.00 from his Dad to buy a used car. He agreed to pay his Dad back in one year with 6% simple interest. How much interest will Michael pay?

    A. $ 27.00
    B. $ 24.00
    C. $270.00
    D. $240.00

29. At Washington High School, 60% of the students ride the bus. What fraction of the students ride the bus to school?

    E. $\frac{2}{3}$

    F. $\frac{1}{6}$

    G. $\frac{3}{8}$

    H. $\frac{3}{5}$

30. In April, A+ Accountants made a $450,000 profit. For the entire year, they had a $750,000 profit. What is the ratio of April profit to total profit?

   A. $\frac{\$3.00}{\$5.00}$

   B. $\frac{\$3.00}{\$8.00}$

   C. $\frac{\$5.00}{\$3.00}$

   D. $\frac{\$8.00}{\$3.00}$

31. 3^3 is equal to

   E. 3
   F. 9
   G. 27
   H. 81

32. Jenna needed one foot square floor tiles for her bathroom.. Her bathroom is 6 feet by 5 feet. The floor tiles come 11 tiles to a box. About how many boxes does she need?

   A. 2
   B. 3
   C. 4
   D. 5

36. Which set of decimals is in order from <u>GREATEST</u> to <u>LEAST</u>?

   E. .3, .36, .036, .06
   F. .95, .9, .095, .09
   G. .048, .41, .481, .5,
   H. .2, .21, .02, .021

34. 14.2 is the same as

   A. $\frac{142}{100}$

   B. $14\frac{1}{50}$

   C. $14\frac{1}{5}$

   D. $14\frac{1}{10}$

Read the following passage. Then choose the answer that makes the most sense in the blank asked for in question 35.

The Gourmet Dining Club has _____ members who meet monthly at a
₁

different Twin Cities restaurant each time. They meet the _____
₂

Saturday of the month, and reservations are always for _____ .
₃

Members have been dining out together for _____ years.
₄

---

35. Which is the best answer for blank 2?

E. 42
F. 7:00 p.m.
G. 3rd
H. 4

36. Which fraction is greater than $\frac{3}{5}$?

A. $\frac{7}{8}$

B. $\frac{1}{3}$

C. $\frac{4}{15}$

D. $\frac{3}{8}$

37. $3^2 + 4 \times 18 \div 9$

E. 4
F. 14
G. 17
H. 26

38. Which of the following is a true statement about the relationship between −5 and 5?

A. $-5 < 5$
B. $-5 = 5$
C. $-5 > 5$
D. $-5 \geq 5$

39. 4^3 is equivalent to

E. 4
F. 12
G. 43
H. 64

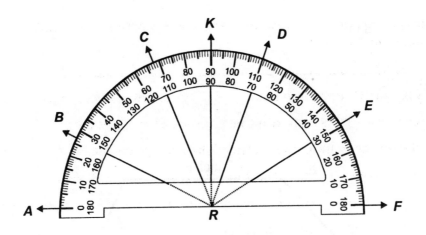

40. What is the measure of ∠ERF?

    A.   31°
    B.   49°
    C.   149°
    D.   151°

41. Use a ruler to find the perimeter of the following rectangle.

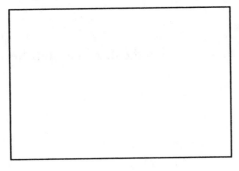

    E.   8 centimeters
    F.   10 centimeters
    G.   20 centimeters
    H.   24 centimeters

42. Which of the following is equal to 300 kilograms?

    A.   0.3 grams
    B.   0.0003 milligrams
    C.   300,000 grams
    D.   30,000,000 milligrams

43. $2\frac{1}{4}$ yards is equal to:

   E.   27 inches
   F.   81 inches
   G.   72 inches
   H.   18 inches

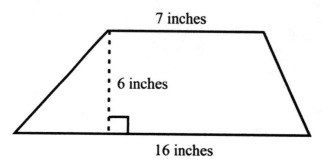

7 inches

6 inches

16 inches

44. Find the area of the trapezoid above.

   A.   42 square inches
   B.   69 square inches
   C.   96 square inches
   D.   672 square inches

45. What is an appropriate measure to use to show the distance from New York City to Miami, Florida?

   E.   kilometers
   F.   meters
   G.   grams
   H.   liters

46. Cynthia added together all the timed laps she made in the pool this week. The total came to 4 hours 62 minutes and 93 seconds. If she simplified the total correctly, it would be equivalent to

   A.   4 hours 2 minutes 33 seconds
   B.   4 hours 3 minutes 33 seconds
   C.   5 hours 3 minutes 33 seconds
   D.   5 hours 4 minutes 33 seconds

47. What is the surface area of a cube that is 2 inches on each side?

   E.   4 square inches
   F.   8 square inches
   G.   16 square inches
   H.   24 square inches

radius = 3 feet

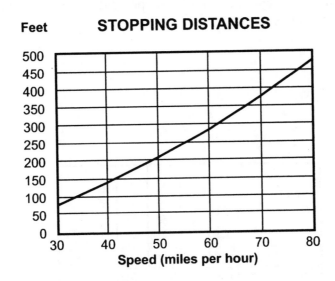

48. Use the formula for volume of a sphere to determine the volume of the hemisphere above.
Volume of a sphere = $\frac{4}{3}\pi r^3$.

A. 4π
B. 12π
C. 18π
D. 36π

Feet   **STOPPING DISTANCES**

49. According to the chart above, how many more feet does it take to stop a car traveling at 70 miles per hour than at 55 miles per hour?

E. 175 feet
F. 150 feet
G. 125 feet
H. 100 feet

LEVELS OF EDUCATION
Persons 25 to 64 Years Old
Data taken from the Statistical Abstract of the United States, 1994

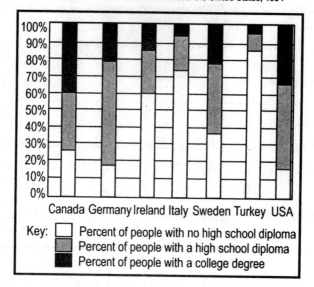

Key:
- ☐ Percent of people with no high school diploma
- ▨ Percent of people with a high school diploma
- ■ Percent of people with a college degree

50. According to the graph above, approximately what percent of people in the USA had a high school diploma but did not have a college degree?

    A.   15%
    B.   50%
    C.   65%
    D.   75%

	Duluth	Minneapolis	Moorhead	Rochester	St. Cloud	St. Paul
Duluth	0	156	253	232	142	150
Minneapolis	156	0	237	90	72	10
Moorhead	253	237	0	328	173	246
Rochester	232	90	238	0	163	82
St. Cloud	142	72	173	163	0	81
St. Paul	150	10	246	82	81	0

51. According to the mileage chart above, how far is Rochester from Minneapolis?

    E.   90 miles
    F.   173 miles
    G.   237 miles
    H.   328 miles

Rubberized Radial Tire Sale		
Item	**Warranty**	**Price**
15-inch radial tire	2 year	$39.00
	5 year	$46.00
16-inch radial tire	2 year	$45.00
	5 year	$52.50
17-inch radial tire	2 year	$54.00
	5 year	$60.50
18-inch radial tire	2 year	$63.00
	5 year	$70.00

52. From the table above, find the price for buying one 16-inch, one 17-inch, and one 18-inch tire each with 2 year warranties.

    A. $162.00
    B. $229.00
    C. $ 63.00
    D. $201.00

LANE PUBLISHING COMPANY
1996 Total Sales = $1,000,000

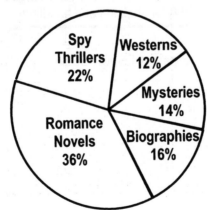

53. According to the pie graph, how much of the 1996 total sales came from Romance Novels?

    E. $120,000
    F. $160,000
    G. $220,000
    H. $360,000

Tax Table - Single Person			
Wages		Exemptions	
At least	But less than	0	1
$136.00	$176.00	$1.30	$0.18
176.00	216.00	2.36	0.85
216.00	256.00	3.61	1.69
256.00	296.00	5.21	2.89
296.00	335.00	6.97	4.27

54. According to the tax table above, how much total tax should be withheld for the following 3 employees: Dan who earned $278.87 and has 0 exemptions, Sue who earned $212.14 and has 1 exemption, and Diane who earned $157.65 and has 1 exemption?

    A. $6.24
    B. $7.75
    C. $8.00
    D. $8.87

SOCKS-A-PLENTY WAREHOUSE		
Item	3 pair price	6 pair price
Anklets	$4.50	$ 8.50
Sport Socks	$6.50	$11.50
Support Socks	$9.00	$15.75
Dress Socks	$6.00	$11.25
Please add 10% for shipping and handling		

55. What would be the total cost of 18 pairs of Sport Socks and 12 pairs of Anklets?

    E. $ 56.65
    F. $ 51.50
    G. $ 57.00
    H. $339.90

A restaurant that serves breakfast between 7:00 a.m. and 11:00 a.m. recorded the times of the morning each customer was served. The histogram below gives data for a typical week.

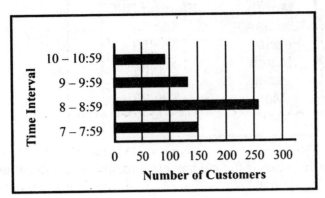

56. During which time interval will customers most likely experience the longest wait for a table?

    A. 7 – 7:59
    B. 8 – 8:59
    C. 9 – 9:59
    D. 10 – 10:59

57. If Charles spins the spinner pictured to the right, which of the following is the **least** likely to happen?

    E. It will land on an even number.
    F. It will land on an odd shaded number.
    G. It will land on an even shaded number.
    H. It will land on a shaded number.

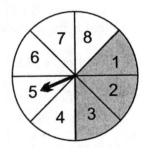

Harold	17 years old
Terry	15 years old
Brenda	12 years old
Colby	11 years old
Alex	10 years old

58. The list above shows the ages of John and Mary's children. What is the average age of the children?

    A. 13
    B. 12
    C. 11
    D. 5

DOBBINS FAMILY

Month	Electric Bill
January	$89.15
February	$99.59
March	$78.99
April	$72.47
May	$99.23
June	$124.69

59. Look at the chart above. What is the median electric bill for the Dobbins family from January through June?

    E. $ 55.22
    F. $ 94.02
    G. $ 94.19
    H. $124.69

60. Darin is playing a dart game at the county fair. At the booth, there is a spinning board completely filled with different colors of balloons. There are 6 green, 4 burgundy, 5 pink, 3 silver, and 8 white balloons. Darin aims at the board with his dart and pops one balloon. What is the probability that the balloon popped is **not** green?

    A.   1 out of 13
    B.   10 out of 13
    C.   3 out of 13
    D.   2 out of 13

61. Matthew has 16 fish in an aquarium. The fish are the following colors: 4 blue, 6 orange, 2 black and white striped, and 4 pink. Matthew also has a trouble-making cat that has caught a fish. What is the probability that the cat will catch an orange fish if all the fish are equally capable of avoiding the cat?

    E. $\frac{1}{4}$

    F. $\frac{3}{8}$

    G. $\frac{1}{8}$

    H. $\frac{1}{6}$

62. Sarah deposits 50¢ into Miss Clucky, a machine which makes chicken squawks and gives Sarah one plastic egg with a toy surprise. In the machine, 30 eggs contain a rubber frog, 43 eggs contain a plastic ring, 23 eggs contain a necklace, and 18 eggs contain a plastic car. What is the probability that Miss Clucky will give Sarah a necklace in her egg?

    A.   1 out of 115
    B.   23 out of 114
    C.   4 out of 5
    D.   1 out of 5

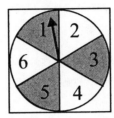

63. What is the probability that the spinner above will land on a shaded section or the number 4?

    E. $\frac{1}{6}$

    F. $\frac{1}{3}$

    G. $\frac{1}{2}$

    H. $\frac{2}{3}$

64. Which of the following is an acute angle?

    A.

    B.   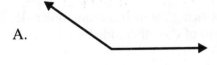

    C.   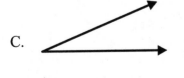

    D.

65. Which of the following objects best represents a sphere?

E.

F.

G.

H.

66. Crystal has a rectangular sheet of wrapping paper measuring 30 inches along one side. The perimeter of the paper is 132 inches. What is the length of the other side?

    A.   30 inches
    B.   36 inches
    C.   72 inches
    D.   102 inches

67. The living room in Ty's house has 168 square feet of floor space. His family is building on an addition to the room that measures 14 feet long and 8 feet wide. What will be the total square feet of the living room with the new addition?

    E.   112 square feet
    F.   180 square feet
    G.   270 square feet
    H.   280 square feet

68. What is the volume of the following box?

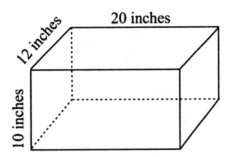

    A.   240 square inches
    B.   240 cubic inches
    C.   2,400 square inches
    D.   2,400 cubic inches

69. Find the area of the following trapezoid.

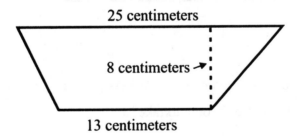

    E.   200 square centimeters
    F.   104 square centimeters
    G.   152 square centimeters
    H.   304 square centimeters

70. A Ferris wheel has a radius of 14 feet. How far will you travel if you take a ride that goes around six times. Use $\pi = \frac{22}{7}$.

    A.   264 feet
    B.   528 feet
    C.   3,696 feet
    D.   12,936 feet

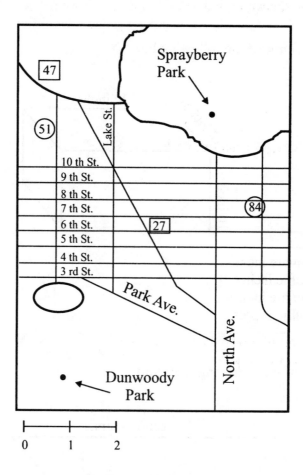

71. If you were at the corner of North Avenue and Park Avenue and needed to go to 10th Street and ⑤⑴, what would be the shortest route to take?

    E.   Park Ave. to 3rd St., over to ⑤⑴, and up to 10th St.

    F.   Park Ave. to Lake St. to 10th St and over to ⑤⑴.

    G.   North Ave. to 3rd St., over to ⑤⑴, and up to 10th.

    H.   North Ave. to 10th St. and over to ⑤⑴.

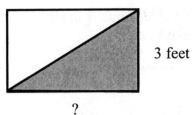

3 feet

?

72. The area of the shaded region of the rectangle above is 6 square feet. What is the length of the rectangle?

    A. 2 feet
    B. 3 feet
    C. 4 feet
    D. 6 feet

73. At the school store, 5 pens sell for $1.25. Which proportion below will help you find the cost of 12 pens?

    E. $\frac{5}{12} = \frac{\$1.25}{x}$

    F. $\frac{5}{12} = \frac{x}{\$1.25}$

    G. $\frac{5}{x} = \frac{12}{\$1.25}$

    H. $\frac{x}{5} = \frac{12}{\$1.25}$

74. There are 200 birds at the **My Feathered Friends** pet store. 128 of the birds are imported from South America, and the rest are domestic. What percent of the birds are domestic?

    A. 72%
    B. 36%
    C. 64%
    D. 28%

75. $24 - 10 \div (3 - 5) + 4^2 =$

    E. 23
    F. 25
    G. 35
    H. 45

Office Mart had a sale on computers in all 21 of its stores. Below is a chart showing how many of each brand were sold.

Wong

Techware

Futura

AAA Computers

Micronet

Formatron

KEY ▯ = 20 COMPUTERS

76. How many brands sold more than 80 computers?

    A. 3
    B. 4
    C. 5
    D. 6

This is the end of the test.